# Properties and
# Products of Algae

# Properties and Products of Algae

Proceedings of the Symposium on the Culture of Algae sponsored by the Division of Microbial Chemistry and Technology of the American Chemical Society, held in New York City, September 7-12, 1969

## Edited by J. E. Zajic

*Biochemical Engineering*
*Faculty of Engineering Science*
*University of Western Ontario*
*London, Ontario, Canada*

 PLENUM PRESS • NEW YORK-LONDON • 1970

*Library of Congress Catalog Card Number 75-112588*

SBN 306-30471-6

© 1970 Plenum Press, New York
A Division of Plenum Publishing Corporation
227 West 17th Street, New York, N. Y. 10011

United Kingdom edition published by Plenum Press, London
A Division of Plenum Publishing Corporation, Ltd.
Donington House, 30 Norfolk Street, London W.C.2, England

Printed in the United States of America

PREFACE


The Fermentation and Biotechnology Division
of the American Chemical Society annually
organizes symposia on topics vital to the
applied biosciences. In September, 1969, a
symposium was held on "Properties and Products
of Algae". Papers presented at this symposium
covered numerous aspects of algal culture with
emphasis on the special properties and products
produced by algae of economic interest. The
art of handling algae is rapidly becoming a
science. The papers presented herein are part
of man's attempt to understand the contribution
of the algae to man's world.

Contributors to this Volume:

H. H. Blecker
Department of Chemistry
University of Michigan Flint College
Flint, Michigan

Y. S. Chiu
Biochemical Engineering
Faculty of Engineering Science
University of Western Ontario
London, Ontario

N. L. Clesceri
Rensselaer Polytechnic Institute
Troy, New York

Thomas H. Haines
The City College of the City University of New York
New York, New York

Raymond W. Holton
Department of Botany
University of Tennessee
Knoxville, Tennessee

Eva Knettig
Biochemical Engineering
Faculty of Engineering Science
University of Western Ontario
London, Ontario

P. J. Lavin
Albany County Health Department
Albany, New York

G. C. McDonald
Albany County Sewer District
Albany, New York

Edward J. Schantz
Department of the Army
Fort Detrick, Maryland

R. D. Spear
Rensselaer Polytechnic Institute
Troy, New York

B. Volesky
Biochemical Engineering
Faculty of Engineering Science
University of Western Ontario
London, Ontario

Varley E. Wiedeman
Department of Biology
University of Louisville
Louisville, Kentucky

J. E. Zajic
Biochemical Engineering
Faculty of Engineering Science
University of Western Ontario
London, Ontario

# CONTENTS

HETEROTROPHIC CULTURE OF ALGAE

J.E. Zajic and Y.S. Chiu

Biochemical Engineering, Faculty of Engineering

Science, University of Western Ontario, London

## INTRODUCTION

Classically all algae form their cellular carbon solely from carbon dioxide by photosynthesis. However, some are facultative heterotrophs and are able to utilize organic substrates as a source of carbon. Also there are obligate heterotrophic algae which must obtain at least some organic compounds from their surroundings. For example, the colorless alga Prototheca zopfii, which does not have the ability to photosynthesize, is unable to grow in the absence of organic materials (Barker, 1935). This alga utilizes ammonia, nitrogen from yeast autolyzate as well as glucose. It is unable to grow in the complete absence of yeast autolyzate.

Under heterotrophic conditions organic nitrogen is not always necessary, however algal growth is often accelerated by a simultaneous provision of both organic carbon and organic nitrogen sources (Kiyohara et al, 1960; Kathrein, 1964; Griffiths, 1967). Kiyohara et al (1960) observed maximum growth of the blue-green alga Tolypothrix tenuis when casamino acids and glucose were added together.

As have been shown by many workers (Iggena, 1938; Pearsall et al, 1940; Neish, 1951; Killam et al, 1956; Samejima et al, 1958; Mineeva, 1962b; Dvorakova-Hladka, 1966; Karlander et al, 1966), heterotrophic growth is accelerated upon introduction of light. For convenience of nomenclature, this type of growth has been called photoheterotrophy or mixotrophy in contrast to photoheterotrophy in which cellular material is synthesized solely from inorganic matter in light. Thus in both photoheterotrophy and

1

photoautotrophy light is involved.  There also exists another type
of nutrition in which light is involved.  That was shown by Algeus
(1948b) in which <u>Chlorella vulgaris</u> grew by using atmospheric
carbon dioxide and glycocoll.  <u>C. vulgaris</u> utilizes the amino
moiety of glycocoll by deamination and by releasing ammonia which
is then assimilated.  This occurs in both light and dark, however,
conversion is much greater in light.  Glycocoll is not a source of
carbon but it does provide a source of organic nitrogen which makes
this process also heterotrophic.  Thus glycocoll is not used as a
source of energy.

Apart from the growth of algae in nature, photoautotrophic
algae production has not been successful on a large scale.  Natural processes in which large scale photoautotrophic culture is conducted are (1) for increasing field fertility (Watanabe, 1962) and
(2) for food production (Nakamura, 1961) and (3) in sewage treatment (Oswald et al, 1957).  All of these are aimed at higher productivity of cellular material itself.  For obtaining maximum cell
concentration in the shortest time high growth rates are required.
In photoautotrophy, use of thin aqueous layers of suspensions is
required in order that light will be effective in reaching the
entire culture.  Such systems require large surface areas which
are not always spatially economic, although this may be partially
overcome by good mixing.  In heterotrophy, even if light is provided, it is not the sole source of energy, thus it is easier to
achieve maximum cell contact with both energy sources with a net
result in greater productivity.

The approach to heterotrophic culture of algae for practical
interest is still relatively new.  The prime objective is to
increase productivity by using cheap substrates such as plant and
animal residues or industrial organic wastes.  Very little information is available on commercial processes.  For example, the
processes involved in producing xanthophylls are still under the
veil of patents and production information is lacking.

Whether photoheterotrophic culture of algae will contribute
to the space program is not known but as man probes further into
space, it can be expected to increase.  Algae which are able to
oxidize or reduce organic matters depending upon the energy source
may be used for (1) to remove the organic materials from waste,
(2) to remove and produce $CO_2$ and $O_2$ and (3) as a source of food.


Algae Able to Grow Heterotrophically

Algal isolates which have been most intensively studied for
heterotrophic culture are those of the green algae such as
<u>Chlorella</u> and <u>Scenedesmus</u>, blue-green algae and other miscellaneous
cultures of algae.

Chlorella. Even in the early work of Beijerinck (1898), it was known that Chlorella could be grown in the dark with glucose as the organic substrate. In 1927, Emerson observed Chlorella cells growing in a medium containing glucose. Under these conditions cells possessed a respiration rate about four times that of light grown cells. More recently oxidative assimilation has been observed for C. pyrenoidosa on acetate and glucose (Myers, 1947; Myers et al, 1949); on glucose, galactose and acetate (Samejima et al, 1958); and for C. ellipsoidea on acetate (Fujita, 1959). Cells grown in dark assimilated about 49% of the glucose carbon (Myers et al, 1949) while Samejima et al (1958) reported the carbon assimilation was 45% for glucose, 37% for galactose and 26% for acetate. In another species C. ellipsoidea, Fujita reported carbon assimilation to be as high as 80% for acetate.

In studies of the heterotrophic nutrition of C. vulgaris Griffiths (1960) reported much higher rates of growth and respiration on glucose than with other organic substrates.

In the studies of xanthophyl production by C. pyrenoidosa, Theriault (1965) reported glucose, fructose and galactose were readily assimilated.

Glycocoll, which has no value as carbon source for Chlorella, is a good source of nitrogen for C. vulgaris (Algeus, 1948b). Utilization is preceded by deamination and release of ammonia to the medium. Further metabolism of the ammonia is quite rapid in the presence of carbon dioxide and light.

In a recent study on organic nitrogen utilization by Griffiths (1967) peptone was found more effective than nitrate for heterotrophic culture of C. vulgaris (Emerson strain). It allows the production of an increased amount of cellular material and supports an enhanced rate of cell division.

Scenedesmus. In the studies of oxidative assimilation, Taylor (1950) found that 80-90% of the glucose carbon was assimilated by S. quadricauda. Mannose is also utilized, but is an inferior substrate compared to glucose. S. obliquus utilizes glucose, cellobiose and acetate. Glucose added in the presence of light was found to be most satisfactory for growth (Dvorakova-Hladka, 1966). As a source of organic nitrogen, glycocoll was the most satisfactory substrate tested for S. obliquus (Algeus, 1948a). Alanine was inferior to glycocoll (Algeus, 1949).

Other green algae. Prototheca, the colorless counterpart of Chlorella, has been studied by Barker (1935, 1936) and Anderson (1945). Barker reported P. zopfii is unable to grow in the absence of an organic nitrogen source such as yeast autolyzate. For carbon sources, conventional monosaccharides, lower fatty acids and simple

alcohols were most commonly assimilated.  It was concluded that
Prototheca is characterized by its ability to assimilate simple
organic molecules.  For example 75% ethanol, and over 80% glycerin
were assimilated.  Of the Krebs-cycle acids none was found to be
utilized.  However, Anderson (1945) reported pyruvate and lactate
were assimilated under oxidative conditions, and oxidation was
stimulated by addition of thiamin.

Algeus (1948c), studying the deamination of glycocoll by green
algae, divided species of the algae into the Scenedesmus and the
Chlorella types.  The former included Haematococcus, Ankistrodesmus
and one species of Hormidium.  These all possessed deaminases.
Algeus suggested that the self-purification of natural waters by
deamination and free ammonia production was performed by this type
of alga.

In the studies of the ability of the green algae Chlorococcum
macrostigmatum to utilize the sugars present in spent sulfite
liquor, Maloney (1959) reported that sugars serve as both carbon
and energy sources for growth in dark.

Studies were also extended to the algae inhabiting sewage,
in which Eppley et al (1962) isolated two strains of Chlamydomonas
from sewage lagoons, and found them capable of utilizing acetate
for $CO_2$ reduction in light without any net production of oxygen.
The two strains were C. mundana Gerloff var. astigmata nov. var.
(Mojave strain) and C. mundana Gerloff (Boron strain).  Wiedeman
et al (1965) grew 39 isolates from waste-stabilization ponds heterotrophically under various culture conditions.  Isolates included
the species of Chlamydomonas, Chlorococcum, Dictyosphaerium,
Pediastrum, Ankistrodesmus etc.  Glucose, mannose and coconut milk
support growth of some of them in the dark with or without air.
In the light, fructose and acetate were also utilized.

Blue-green algae.  As suggested by Fogg in 1956, the frequent
occurrence of blue-green algae in soils and habitats rich in organic matters indicated that they grew in dark on organic matter.

Earlier Harder (1917) reported Nostoc punctiforme was capable
of growing on carbohydrates.  Allison et al (1937) observed a
strain of Nostoc muscorum which grew and fixed nitrogen on glucose
in the dark.  But it has not been until recently that more detailed
studies on heterotrophic culture have involved the blue-green algae.

In examining the conditions favoring heterotrophic growth of
Tolypothrix tenuis, Kiyohara et al (1960) reported maximal growth
on addition of both casamino acids and glucose.  Certain specific
amino acids, such as arginine and phenylalanine, were also utilized
as a source of nitrogen.

In studies concerning nitrogen fixation which is common among blue-green algae, Fay (1965) reported sucrose was most readily utilized by Chlorogloea fritschii. The growth rate of C. fritschii increased fourfold in light and sucrose assimilation was greater in the light in the absence of $CO_2$. Another heterotrophic nitrogen fixation alga Anabaenopsis circularis, utilizes glucose, fructose and sucrose (Watanabe et al, 1967). The effect of light was examined with fructose and it stimulated growth.

Diatoms. In 42 cultures of diatoms from freshwater and soil, Lewin (1953) found only 13 capable of growing on glucose. The 13 cultures comprised seven isolates of Navicula pelliculosa, five of other species of Navicula, and one unidentified isolate. One isolate of Navicula pelliculosa was tested for its ability to grow heterotrophically. Sixty common carbohydrates, fatty acids, alcohols and amino acids were tested. Only glucose was utilized. In the presence of light, glycerol and fructose also permitted growth.

In a later study, Lewin et al (1960) examined 44 isolates of marine littoral diatoms and found more than half consisting of species of Amphora, Navicula, Nitzschia and Cyclotella were able to grow heterotrophically. Glucose was the preferred carbohydrate.

Though there are still other divisions, genera and species of algae which grow heterotrophically, the above mentioned cultures are of the most significant. There are in excess of 88 literature citations on algae growing under heterotrophic and mixotrophic conditions. These have been summarized in Table I. The algal culture(s) used, general classification, substrates utilized, growth conditions and references are summarized.


Effect of Light on Heterotrophic Growth

Many studies comparing heterotrophic growth in light and in dark have been reported. The best conclusion is that light stimulates growth of algae under heterotrophic culture.

During photosynthesis carbon dioxide is reduced and converted into cellular material by absorbing light energy, whereas in fact photosynthetic organisms grow by utilizing the chemical energy which is wholly or partly produced during light absorption. As stated previously, light stimulates growth during photoheterotrophy, therefore this type of growth is, in a sense, photosynthetic in which organic substrates are photoassimilated. For example, in higher plants Maclachlan et al (1959) used tobacco leaf disks to synthesize starch and sucrose from glucose in light under anaerobic conditions. It was concluded that light replaced the oxygen requirement for synthesis by inducing an anaerobic phosphorylation.

Table 1. Organic substrates and growth conditions of heterotrophic algae

| Organism (Classification) | Organic C and/or N Source | Growth Conditions | Reference (Remarks) |
|---|---|---|---|
| Anabaenopsis circularis (2) | glucose** fructose* sucrose | 0.5% 0.3%; dark/light* 0.3% | Watanabe et al 1967 (nitrogen fixation) |
| Ankistrodesmus braunii (1) | glucose | | Bishop 1961 |
| A. falcatus | glucose + glycocoll** glycocoll* alanine | dark/light* dark/light* dark/light* | Algeus 1948c, 1949 |
| Botrydiopsis intercedens (5) | glucose acetate | | Belcher et al 1960 |
| Bracteacoccus cinnabarinus (1) (Kol et F. Chod.) Starr | glucose acetate | | Parker et al 1961 |

Table 1. (Cont..)

| Organism | Substrate | Reference |
|---|---|---|
| B. engadiensis (Kol et F. Chod.) Starr | glucose | Parker et al 1961 |
| B. minor (Chodat) Petrova | glucose | Parker et al 1961 |
| B. terrestris (Kol et F. Chod.) Starr | glucose | Parker et al 1961 |
| Bracteacoccus sp. | glucose | Parker et al 1961 |
| Bracteacoccus sp. | glucose acetate | Parker et al 1961 |
| Bumilleriopsis brevis (5) | glucose* mannose fructose* sucrose acetate* citrate fumarate succinate glycine glycerol ethanol | Belcher et al 1960 |

Table 1. (Cont..)

| | | | |
|---|---|---|---|
| Chilomonas paramecium (6) | acetate<br>lactate | thiamin<br>thiamin | Hutchens 1940, 1948<br>Hutchens et al 1948 |
| | ethanol | 0.01 M | Pace, 1948 |
| | acetate<br>ethanol | | Blum et al 1951 |
| | ethanol<br>1-butanol<br>1-hexanol<br>acetate<br>n-butyrate<br>n-caproate<br>n-caprylate | | Cosgrove 1950<br>Cosgrove et al 1952 |
| Chlamydobotrys sp. (1) | acetate + tryptone<br>+ yeast extract<br>+ beef extract | light only<br>2000-5000 lux<br>25-30°C | Pringsheim et al 1960 |
| | acetate | light | Wiessner et al 1964 |
| Chlamydomonas agloeformis (1) | glucose + CSL ‡<br>+ urea | | Kathrein 1960<br>(xanthophylls) |
| C. dysosmos | acetate**<br>pyruvate<br>lactate | dark/light* | Lewin 1954 |

Table 1. (Cont..)

| | | | |
|---|---|---|---|
| C. mundana<br><br>Mojave strain<br>Boron strain | acetate | light only<br>NH$_3$<br><br>29-34°C<br>28-32°C | Eppley et al 1962,<br>1963a |
| C. reinhardi<br>(-) strain | acetate | light | Eppley et al 1962 |
| Chlorella<br>ellipsoidea (1) | glucose<br>glucose + formate*<br>glucose + urea**<br>galactose | dark/light*<br><br><br>dark/light* | Samejima et al 1958 |
| | acetate | | Fujita 1959 |
| C. prototheocoides | glucose + glycocoll<br>glucose + alanine**<br>glycocoll* | dark/light*<br>dark*/light<br>dark/light* | Algeus 1948c, 1949 |
| C. pyrenoidosa | glucose | dark/light* | Bergmann 1955 |
| | acetate | dark/light* | Schlegel 1956 |
| | glucose**<br>galactose*<br>acetate | dark/light*<br>dark/light* | Samejima et al 1958 |

Table 1. (Cont..)

| | | dark/light | Dvorakova–Hladka 1966 |
|---|---|---|---|
| C. pyrenoidosa ATCC 7516 | glucose galactose** acetate | 1080/1800 lux 1080*/1800 lux 1080*/1800 lux | |
| C. pyrenoidosa C-37-2 | glucose + CSL + urea | | Kathrein 1960 (xanthophylls) |
| | glucose + tryptone + yeast extract | | Guehler et al 1962 (steroid) |
| C. pyrenoidosa Emerson | glucose | dark/light* | Killam et al 1956 McBride 1961 |
| C. pyrenoidosa 7-11-05 | glucose + urea** glycerol + urea* dextrin + urea | dark-batch/ light-batch*/ light-continuous** | Theriault 1965 (xanthophylls) |
| | glucose + CSL + urea | | Zajic 1969 (xanthophylls) |
| C. vulgaris | glucose + glycocoll** glucose + alanine glycocoll | dark/light* dark/light dark/light* | Algeus 1948b, 1949 |

Table 1. (Cont..)

| | Substrate | Conditions | Reference |
|---|---|---|---|
| | glucose | dark/light* | Finkle et al 1950 <br> Killam et al 1956 <br> McBride 1961 |
| | glucose* <br> acetate | dark/light* <br> 2000–5000 lux <br> 700–1000 lux | Mineeva 1962a, 1962b |
| C. vulgaris <br> ATCC 9765 | glucose + CSL <br> + urea | | Kathrein 1960 <br> (xanthophylls) |
| C. vulgaris <br> Emerson | glucose | dark/light* | Karlander et al 1966 |
| | glucose* <br> glucose + peptone** <br> sucrose | | Griffiths 1960, 1967 <br> (protein) |
| C. vulgaris <br> Pratt–Trealease | glucose** <br> galactose <br> fructose* <br> lactose <br> cellobiose <br> aseculin <br> methyl-β-D-glucoside | 35°C <br> dark/light* <br> dark/light* <br> dark/light* <br> dark/light* <br> dark/light* <br> dark/light* <br> dark*/light | Neish 1951 |

Table 1. (Cont...)

| Organism | Substrate | Conditions | Reference |
|---|---|---|---|
| C. vulgaris var. viridis | glucose + acetate + glycine + yeast extract | thiamin | Pruess et al 1954 |
| Chlorellidium tetrabotrys (5) | acetate citrate succinate glycine glycerol ethanol | | Belcher et al 1960 |
| Chlorocloster engadinensis (5) Vischer | glucose galactose** mannose fructose | dark/light* 200-300 lux 23°C | Clemencon 1965 |
| Chlorococcum macrostigmatum (1) Starr | spent sulfite liquor (sugar) | dark/light* | Maloney 1959 |
| Chlorococcum sp. | glucose + CSL + urea | | Kathrein 1960 (xanthophylls) |
| Chlorogloea fritschii (2) | sucrose** | dark/light* 0.01 M | Fay et al 1962 Fay 1965 (nitrogen fixation) |

Table 1. (Cont..)

| | | | |
|---|---|---|---|
| | maltose glycine glutamine aspartate | | |
| Chlorogonium elongatum (1) | sucrose | 35°C | Holton 1968 |
| | acetate | light | Pringsheim et al 1960 |
| Coccomyxa elongata (1) | glucose + CSL + urea | | Kathrein 1963 (xanthophylls) |
| Cyclotella sp. (3) | glucose | vitamin $B_{12}$ | Lewin et al 1960 |
| Dictyochloris fragrans (1) Vischer | glucose | | Parker et al 1961 |
| Euglena gracilis (7) | acetate | light | Pringsheim et al 1960 |
| E. gracilis Vischer | acetate butyrate* | pH 6.8 6.8 | Cramer et al 1952 |
| E. gracilis var. bacillaris | | thiamin vitamin $B_{12}$ | Cramer et al 1952 |

Table 1. (Cont..)

| | | | |
|---|---|---|---|
| | glucose* | pH 4.5 | |
| | acetate | 6.8 | |
| | pyruvate | 4.5 | |
| | succinate** | 4.5 | |
| | fumarate | 4.5 | |
| | butyrate | 6.8 | |
| | glutamate | 4.5 | |
| | alanine | 4.5 | |
| | aspartate | 4.5 | |
| | acetate | dark/light* | Lynch et al 1953 |
| | acetate | | Wilson et al 1958 |
| E. gracilis var. saccharophila | glucose | | Pringsheim 1955 |
| | fructose | | |
| Haematococcus pluvialis (1) | glycocoll | light only | Algeus 1948c |
| | glucose + glycocoll | | |
| Navicula incerta Grun. (3) | glucose | | Lewin et al 1960 |
| N. pelliculosa | glucose** | dark/light* | Lewin 1953 |
| | fructose | light only | |
| | glycerol | light only | |
| Neochloris alveolaris (1) Bold | glucose | | Parker et al 1961 |
| | acetate | | |

Table 1. (Cont..)

| Species | Substrate | Requirement | Reference |
|---|---|---|---|
| N. aquatica Starr | glucose | | Parker et al 1961 |
| N. gelatinosa Herndon | glucose | | Parker et al 1961 |
| N. pseudoalveolaris Deason et Bold | glucose | | Parker et al 1961 |
| Neochloris sp. | glucose | | Parker et al 1961 |
| Nitzschia angularis var. affinis (3) (Grun.) perag. | glucose | | Lewin et al 1960 |
| N. closterium (Ehr.) W. Sm. | glucose lactate | thiamin | Lewin et al 1960 |
| N. curvilineata Hust. | lactate | | Lewin et al 1960 |
| N. filiformis W. Sm. | glucose | | Lewin et al 1960 |
| N. frustulum (Kürtz.) Grun. | glucose lactate | vitamin $B_{12}$ | Lewin et al 1960 |

Table 1. (Cont..)

| | | | |
|---|---|---|---|
| N. laevis Hust. | glucose lactate | | Lewin et al 1960 |
| Nostoc muscorum (2) | glucose sucrose | dark/light* dark/light* | Allison et al 1937 |
| | acetate | dark/light* | Allison et al 1953 |
| Ochromonas malhamensis (4) | glucose | thiamin biotin vitamin $B_{12}$ | Reazin 1956 |
| Pediastrum boryanum (1) | glucose** mannose acetate coconut milk | dark/light* light only light only | Wiedeman et al 1965 |
| P. duplex | glucose** mannose acetate coconut milk | dark/light light only light only | Wiedeman et al 1965 |
| Polytoma obtusum (1) | acetate | thiamin | Lwoff 1947 |
| P. ocellatum | acetate | thiamin only | Lwoff 1947 |
| P. uvella | acetate | thiamin | Lwoff 1947 |

Table 1. (Cont..)

| | | | |
|---|---|---|---|
| Polytomella caeca (or coeca) (1) | acetate | thiamin only | Lwoff 1947 |
| | acetate | | Barker et al 1955 |
| Prototheca zopfii (1) | yeast autolyzate + <br><br>glucose<br>fructose<br>mannose<br>galactose<br>acetate<br>propionate<br>n-butyrate<br>iso-butyrate<br>n-valerate<br>glycerol<br>ethanol | 30°C | Barker 1935, 1936 |
| | yeast agar + <br><br>pyruvate<br>lactate<br>glycerol | thiamin | Anderson 1945 |
| Scenedesmus acuminatus (1) | glucose +<br>glycocoll*<br>glucose + alanine**<br>glycocoll | dark*/light<br>dark/light<br>dark/light | Algeus 1948c, 1949 |

Table 1. (Cont..)

| Species | Substrate | Conditions | Reference |
|---|---|---|---|
| S. acutiformis | glucose + glycocoll**<br>glucose + alanine<br>glycocoll | dark/light*<br>dark/light*<br>dark/light* | Algeus 1948c, 1949 |
| S. costulatus Chod, var. chlorelloides | glucose | dark/light* | Bristol-Roach 1928 |
| S. dimorphus | glucose + glycocoll**<br>glucose + alanine*<br>glycocoll | dark/light*<br>light only<br>dark/light* | Algeus 1948c, 1949 |
| S. obliquus | glucose | dark/light* | Killam et al 1956 |
|  | glucose*<br>acetate | dark/light*<br>2000-5000 lux | Mineeva 1962a, 1962b |
|  | glucose**<br>cellobiose<br>acetate | 1080 /1800* lux<br>1080 /1800* lux<br>1080*/1800 lux | Dvorakova-Hladka |
|  | glucose + glycocoll**<br>glycocoll*<br>alanine | dark/light*<br>dark/light*<br>dark/light* | Algeus 1948a, 1949 |

Table 1. (Cont..)

| Species | Substrate | dark/light | Reference |
|---|---|---|---|
| S. quadricauda | glucose + glycocoll* glycocoll | dark/light* dark/light* | Algeus 1948c |
|  | glucose* mannose |  | Taylor 1950 |
| Spongiochloris excentrica Starr (1) | glucose |  | Parker et al 1961 |
| S. lamellata Deason et Bold | glucose |  | Parker et al 1961 |
| S. spongiosus Starr | glucose |  | Parker et al 1961 |
| Spongiochloris sp. | glucose + CSL + urea |  | Zajic 1969 (xanthophylls) |
| Spongiococcum alabamense (1) Deason | glucose |  | Parker et al 1961 |
| S. excentricum | glucose + CSL + urea |  | Kathrein 1963 (xanthophylls) |
| S. excentricum Deason et Bold | glucose |  | Parker et al 1961 |

Table 1. (Cont..)

| | | | |
|---|---|---|---|
| S. multinucleatum Deason et Bold | glucose | | Parker et al 1961 |
| Stichococcus bacillaris (1) | glucose + CSL + urea | | Kathrein 1960 (xanthophylls) |
| S. subtilis | glucose + acetate + glycine + yeast extract<br><br>glucose + acetate + enzyme hydrolyzed casein | thiamin | Pruess et al 1954 |
| Tolypothrix tenuis ((2) | glucose + casamino acids**<br><br>glucose + arginine<br>glucose + phenylalanine | glucose 1% casamino acids 0.5% pH 8.0 32°C | Kiyohara et al 1960 |
| Tribonema aequale (5) Pascher | a. agar + glucosé sucrose acetate citrate | pH 5.8 5.8 4.0 4.0 | Belcher et al 1958 |

Table 1. (Cont..)

| | | | Belcher et al 1960 |
|---|---|---|---|
| b. glycine*<br>L-valine*<br>DL-serine<br>DL-phenylalanine<br>DL-tryptophane<br>L-histidine<br>L-glutamine<br>urea*<br>urethane<br>L-leucine<br>DL-threonine<br>L-tyrosine<br>L-cystine*<br>L-arginine-HCl*<br>L-citrulline<br>acetamide<br>succinimide<br>guanidine carbonate | | | |
| _T. minus_ | glucose | | |

(1): Division Chlorophyta (green algae)
(2): Division Cyanophyta (blue-green algae)
(3): Division Bacillariophyta (diatoms)
(4): Division Chrysophyta
(5): Division Xanthophyta
(6): Division Cryptophyta
(7): Division Euglenophyta
*: more satisfactory
**: most satisfactory
#: CSL = corn steep liquor

In other instances certain algae such as <u>Chlorella</u>, <u>Chlamydobotrys</u>, <u>Chlorogonium</u> and <u>Euglena</u> were observed to grow anaerobically in the absence of carbon dioxide when acetate and light were provided (Pringsheim et al, 1960; Wiessner et al, 1964). Whatley et al (1964) concluded that light energy is required only to supply ATP (adenosine triphosphate) by cyclic photophosphorylation.

Assimilation of acetate and its enhancement by light was also investigated in <u>Chlamydomonas mundana</u> by Eppley et al (1963a). They reported that the increase of both acetate assimilation and respiration appears related to photosynthetic production of ATP, and that the inability to grow in dark with acetate is due to limited oxidative phosphorylation. Some individuals question whether this is true heterotrophy, these same individuals should ponder as to whether this is true phototrophy and at this point they will realize that this is an excellent example of photoheterotrophy, particularly if the algae culture can live in the dark on the substrate in question.

Among the earlier workers studying the effect of light on heterotrophic growth are Bristol-Roach (1928), Gaffron et al (1942) for <u>Scenedesmus</u>; Myers (1947), Killam et al (1956) and Samejima et al (1958) for <u>Chlorella</u>. Gaffron et al (1942) reported that <u>Scenedesmus</u>, which had been fermenting for several hours in the dark in a medium containing glucose gave an increase of several fold of free hydrogen upon re-illumination. In studies of oxidative assimilation of glucose by <u>C</u>. <u>pyrenoidosa</u>, Myers (1947) reported the time required for complete utilization of a limited amount of organic substrate is not affected by illumination, particularly in the absence of carbon dioxide. For <u>C</u>. <u>vulgaris</u>, Killam et al (1956) reported that very low light intensity markedly accelerated growth.

In contrast to light effect on heterotrophic growth, the effect of organic matters on the photosynthesis has been studied by Kratz et al (1955) and Pearce et al (1969) for the blue-green alga <u>Anabaena variabilis</u>. Kratz et al reported that even though glucose added in the autotrophic growth medium was incorporated and metabolized, there was no increase in the rate of growth and only a slight increase in the rate of respiration. Also, Pearce et al (1969) reported that the growth rate of <u>A</u>. <u>variabilis</u> was not increased or retarded by the addition of glucose or sucrose. No growth occurred in the complete absence of carbon dioxide and the presence of glucose. The respiration rate was not stimulated by glucose addition. Possibly this type of photosynthetic growth should have a special name other than photoheterotrophy.

Recently, more quantitative studies concerning the effect of light on growth have been conducted by different workers indicating the following: (A) Dvorakova-Hladka (1966) grew <u>Chlorella</u>

pyrenoidosa and Scenedesmus obliquus in a medium with and without
the presence of 0.05 M sugars (fructose, galactose, glucose,
maltose, cellobiose, sucrose and soluble starch), or 0.025 molar
sodium salts of organic acids of acetate, pyruvate and succinate.
Identical heterotrophic conditions were also tested with specified
levels of light intensity. For Chlorella, the best growth was with
galactose in the light. For Scenedesmus galactose became the least
effective among hexoses, with glucose and light being satisfactory.
There was no significant difference between light intensities of
1080 and 1800 lux. Figures 1 and 2 show only examples where growth
is most different. (B) Mineeva (1962a, 1962b) grew Chlorella
vulgaris and Scenedesmus obliquus under conditions of phototrophic,
mixotrophic and heterotrophic nutrition. For heterotrophic growth,
glucose or acetate was added to the inorganic medium. An increase
occurred at low intensities up to 500-700 lux. Above this, the
ratio of growth between the autotrophic and the heterotrophic
systems became constant over a wide range of light intensities.
Of note was the growth stimulating effect at low light intensities
of the range 50-100 lux in the glucose containing medium. This
was especially true for Scenedesmus at 100 lux. At light satura-
tion (2000-10,000 lux) and with 2 mg of glucose, 1 mg of cellular
material was obtained with Chlorella. In Scenedesmus under the
same conditions only 1.26-1.51 mg glucose was required. In the
dark about 3 mg glucose is required for synthesis of one mg of
either Chlorella or Scenedesmus. Results are shown in Figures 3
and 4.

   Though the studies of the effect of light on heterotrophic
growth has been confined to a limited number of species of algae,
it can be concluded that light stimulates growth. As emphasized
by McBride (1961), light of one lux intensity, which is inadequate
to support autotrophic growth, has been reported to markedly
stimulate heterotrophic growth of the Emerson strain of Chlorella
vulgaris. As shown in the preceding Figures, the effect of light
is not merely additive. The total rate of photoheterotrophic
growth is greater than the sum of the rates due to either photoautotrophy or heterotrophy in the dark. Actually organic compounds
cannot be regarded as stimulating photosynthesis unless actual
utilization of the organic compounds is shown


                           Xanthophyll Production

   Xanthophylls, which are the principal pigments in egg yolk
and the shank in broilers, are synthesized by many algae. They
are also the minor carotenoid pigments in many cereal grains, e.g.
corn. Production of carotenoid pigments by heterotrophic culture
of algae has been studied by Claes (1957, 1959), Kathrein (1960,
1963, 1964), Theriault (1965), Farrow et al (1966), and Zajic (1969).

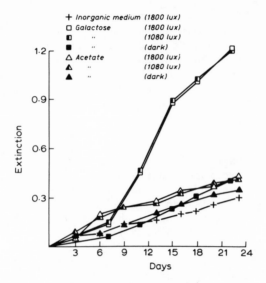

Fig. 1.  Growth of <u>Chlorella</u> <u>pyrenoidosa</u> on various organic substrates.  (After Dvorakova-Hladka, 1966).

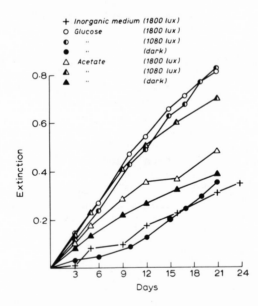

Fig. 2.  Growth of <u>Scenedesmus</u> <u>obliquus</u> on various organic substrates.  (After Dvorakova-Hladka, 1966).

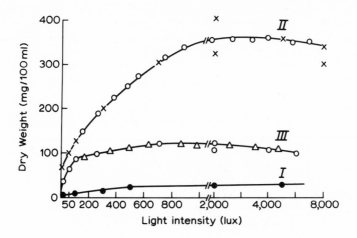

Fig. 3.   Growth of <u>Chlorella</u> <u>vulgaris</u> at different light
intensities.
I   :  Basal medium, phosphate buffer (pH 7.2-7.4)
II  :  Basal medium, plus glucose, phosphate buffer (pH 5.5-6.0)
III:  Basal medium, plus acetate, phosphate buffer (pH 7.1-7.5).
(After Mineeva, 1962b).

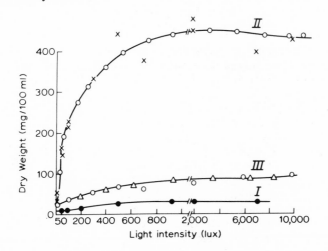

Fig. 4.   Growth of <u>Scenedesmus</u> <u>obliquus</u> at different light
intensities.
I   :  Basal medium, phosphate buffer (pH 7.2-7.4)
II  :  Basal medium, plus glucose, phosphate buffer (pH 5.5-6.0)
III:  Basal medium, plug acetate, phosphate buffer (pH 7.1-7.5)
(After Mineeva, 1962b).

Zajic (1969) grew Chlorella pyrenoidosa 7-11-05, and Spongiochloris sp. on a medium consisting of glucose, corn steep liquor
(CSL) and urea. Average yield of xanthophylls obtained for
Spongiochloris sp. was 350 mg/1. Total ingredient concentrations
were 15-18% glucose, 5-6.5% CSL, and 1.0-1.4% urea. The initial
levels were 3% glucose, 1.5% CSL and 0.4% urea with remaining
amounts fed after 48 hours at intervals of 24-36 hours at the
levels of 3.0, 1.0 and 0.4% respectively.

Kathrein (1960, 1963, 1964) tested many green algae, e.g.
Chlamydomonas agloeformis, Chlorella vulgaris, C. pyrenoidosa,
Chlorococcum sp., Coccomyxa elongata, Spongiococcum excentricum
and Stichoccus bacillaris, for their ability to produce xanthophylls.
The highest yield reported was 294 mg/1 for S. excentricum. The
medium consisted of 3.0% dextrose, 1.5% CSL and 0.4% urea. This
medium was supplemented with 3.0% dextrose and 1.0% CSL at 3, 4,
5, 6 and 7 days, and with 0.4% urea at 5 and 7 days.

Farrow et al (1966) claims to have developed UV mutants of a
Chlorella culture which possessed a temperature growth optima of
38-40°C. One of these mutants gave a lutein yield of 235 mg/1 in
96-120 hours. The medium consisted of 54.6 g glucose per liter
(as glucose monohydrate), 7.5 g invert blackstrap molasses per
liter, 2.4 g ethanol per liter and small amounts of mixed amino
acids, potassium acetate, thiamin, and supplementary salts and
trace elements. Guanidine, supplemented with lysine and several
other amino acids, were particularly good sources of nitrogen.
The addition of 30 mg/1 of thiamin to the medium, along with 3.3
g/1 of guanidine, increased the lutein content of the total
carotenoid mixture from about 78 to about 90%.

In a study by Theriault (1965), excellent production was
realized in 30-liter fermentor when Chlorella pyrenoidosa 7-11-05
was grown on glucose as the sole source of carbon. In 162 hours,
a dry cell weight in excess of 100 g/1 and total xanthophylls of
467-512 mg/1 were obtained from 230-260 g/1 glucose. Using a
continuous feed, a dry cell weight of 302 g/1 and total xanthophyll
level of 650 mg/1 was obtained from 520 g/1 of glucose. Figures
5 and 6, and Table 2 show the results of typical fermentations.
Under batch growth, light stimulated total xanthophyll production.
Under continuous growth in light, total xanthophylls increased
almost twofold, whereas cell weight increased almost threefold over
that obtained heterotrophically in the dark. The greater increment
of cellular production in relation to xanthophylls in continuous
systems in comparison to batch, emphasizes a need for further
investigation of a continuous process.

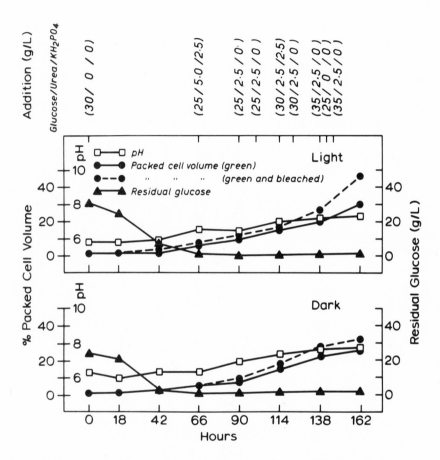

Fig. 5.    Batch growth of <u>Chlorella</u> <u>pyrenoidosa</u> 7-11-05
(Modified from the figures of Theriault, 1965).

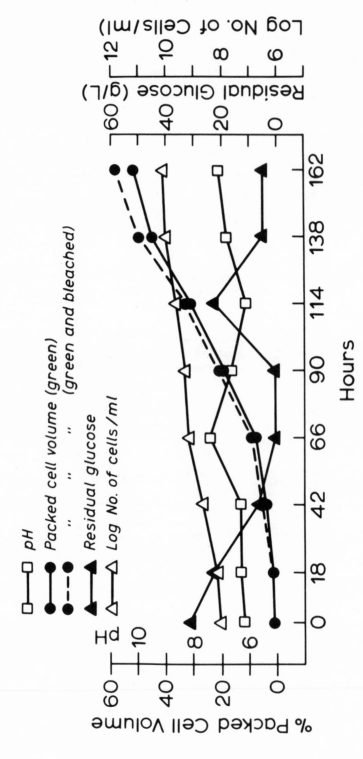

Fig. 6. Growth of <u>Chlorella pyrenoidosa</u> 7-11-05 by continuous feed method. Continuous-feed solution consisting of 605 g/1 glucose, 0.14 g/1 $KH_2PO_4$ and 75 g/1 Seitz-filtered urea was started at 66 hr. The feed rate was increased daily. (After Theriault, 1965).

Table 2. Heterotrophic growth of Chlorella pyrenoidosa 7-11-05: variation of total glucose, dry cell weight and total xanthophylls between batch and continuous method. (Data according to Theriault, 1965).

| Method | Total Glucose (g/1) | Dry Cell Weight 162 hr(g/1) | Total Xanthophylls 162 hr(mg/1) |
|---|---|---|---|
| Batch dark | 230 | 108.6 | 378 |
| light | 260 | 108.8 | 470 |
| Continuous light | 521 | 302.5 | 650 |

Steroid Transformation

The use of steroids in medicine has increased greatly.  Microbial transformation of steroids has been an important contribution.
It was not until recently that algae have been used in effecting
steroid changes.

Luedemann et al (1961) grew Scenedesmus sp. J9A21 under mixotrophic conditions.  4-Pregnene-17α,21-diol-3,20-dione (Reichstein's
Compound S) was added at a level of 100 mg/l to an actively growing
culture of Scenedesmus in a NZ amine (Type A) - starch medium.  The
steroidal product was 4-pregnene-6β,17α,21-triol-3,20-dione.

Guehler et al (1962) grew Chlorella pyrenoidosa C-37-2 also
mixotrophically.  A culture volume of 750 ml composed of 0.75 g
glucose, 3.75 g tryptone and 1.85 g yeast extract was utilized.
After 48 hours of incubation, 0.1875 g 4-androstene-3,17-dione
dissolved in 3.75 ml of 95% ethanol was added.  An 8.5% yield of
testosterone was reported.

In a later study, Guehler (1966) grew 18 species of Scenedesmus,
Chlorella, and Euglena under both autotrophic and mixotrophic conditions.  Dehydroisoandrosterone, 4-androstene-3,17-dione,
progesterone, cholesterol and stigmasterol were individually dissolved in ethanol and separately as substrates for respective
cultures.  With dehydroisoandrosterone as substrate: (1) Scenedesmus
S-2 gave product of 3β,11α-dihydroxy-5-androsten-17-one;  (2) C.
pyrenoidosa C-37-2 gave products of 7α-hydroxydehydroisoandrosterone
and 3β-acetoxy-7α-hydroxy-5-androsten-17-one; and (3) C. ellipsoidea
CE-1 gave a product of 5-androstene-3β,17β-diol with a mixture of
C16, C16:1, C16:2, C16:3, and C18:1, C18:2 and C18:3 acids.  C.
pyrenoidosa C-37-2 converted 4-androstene-3,17-dione to testosterone
and testosterone acetate, whereas a culture adapted to progesterone
and ethanol, transformed progesterone to 5α-pregnane-3,20-dione
and 3β-hydroxy-5α-pregnan-20-one.  4-Androstene-3,17-dione was
converted to androsterone, androstane-3α,17β-diol, and testosterone
by E. gracilis var. bacillaris.  E. gracilis converted progesterone
to 5α-pregnane-3,20-dione, 3α-hydroxy-5α-pregnan-20-one, 20β-
hydroxy-4-pregnen-3-one, 5α-pregnane-3α,20β-diol, and 5α-pregnane-
3β,20β-diol.

Heterotrophic algae which contribute to steroid transformation
and the steroidal products formed are summarized in Table 3.  It
should be noted that rate of steroid transformation is slow (Guehler
1966).

Polysaccharide Accumulation

Starch, which is the most common polysaccharide produced by

Table 3. Steroidal products by heterotrophic algae.

| Organism | Substrate | Product | Reference |
|---|---|---|---|
| Scenedesmus sp. J9A21 | 4-pregnene-17α,21-diol-3,20-dione | 4-pregnene-6β,17α,21-triol-3,20-dione | Leudemann et al 1961 |
| Scenedesmus sp. S-2 | dehydroisoandrosterone | 3β,11α-dihydroxy-5-androsten-17-one | Guehler 1966 |
| Chlorella pyrenoidosa C-37-2 | 4-androstene-3,17-dione | testosterone | Guehler et al 1962 Guehler 1966 |
| | dehydroisoandrosterone | 7α-hydroxydehydro-isoandrosterone | Guehler 1966 |
| | | 3β-acetoxy-7α-hydroxy-5-androsten-17-one | |
| | progesterone | 5α-pregnane-3,20-dione | Guehler 1966 |
| | | 3β-hydroxy-5α-pregnan-20-one | |

Table 3. (Cont..)

| Organism | Substrate | Product | Reference |
|---|---|---|---|
| C. ellipsoidea CE-1 | dehydroisoandro-sterone | 5-androstene-3β,17β-diol | Guehler 1966 |
| Euglena gracilis var. bacillaris | 4-androstene-3,17-dione | androsterone<br>androstane-3α,17β-diol<br>testosterone | Guehler 1966 |
| | progesterone | 5α-pregnane-3,20-dione<br>3α-hydroxy-5α-pregnan-20-one<br>20β-hydroxy-4-pregnen-3-one<br>5α-pregnane-3α,20β-diol<br>5α-pregnane-3β,20β-diol | Guehler 1966 |

higher plants is also produced by algae. Algae vary greatly in
their ability to accumulate starch and starch like compounds.
Bailey et al (1954) reported that 20% of the dry weight of
Chlorella vulgaris grown under certain conditions is starch.

As indicated previously the fraction of carbon assimilation
varies between different algae and the growth conditions used.
Starved cells may assimilate as much as 83% of the organic substrate (Myers, 1947). Normally the greater the percent of carbon
assimilation, the greater the yield of polysaccharide. The polysaccharides synthesized have not been fully characterized. In
Scenedesmus quadricauda (Taylor, 1950) the polymer formed was not
starch, while in Prototheca zopfii (Barker, 1935) and Oscillatoria
princeps (Fredrick, 1951), the polymer was glycogen or glycogen-
like.

The Emerson strain of Chlorella vulgaris (Griffiths, 1963,
1965) synthesizes starch but not the Brannon No. 1 strain. The
two strains were grown under autotrophic and heterotrophic conditions. The cells were starved for 24 hours in an unaerated inorganic medium in dark prior to inoculation. Cells of Emerson
strain accumulated more polymer in dark on glucose than cells
grown autotrophically. Under the same test condition Brannon
No. 1, gave inferior results. Figures 7, 8 and 9 show the effects
of 0.05 M glucose upon the population density, the dry weight and
the average size of cells for the Emerson strain. Results for the
carbohydrate accumulated by the Emerson strain are expressed in
terms of reducing value in Table 4.

It is clear from the data that the increase in average cell
size occurs during the lag phase and is induced by glucose addition
(Fig. 9), whilst the population density increases at a slower
rate in the lag phase under heterotrophic conditions (Fig. 7).
The rates of dry weight increase under the both autotrophic and
heterotrophic conditions appeared similar (Fig. 8).

The relatively low reducing values of alcoholic extracts from
both autotrophic and heterotrophic cultures indicate a low level
of free reducing sugar (Table 4). One assumes the bulk of the
glucose consumed is either (1) oxidized or (2) converted into some
nonreducing substances or to cellular material. In view of the
amount of reducing substance produced following acid hydrolysis of
the cellular extract, there is a marked difference between autotrophic and heterotrophic cultures. The difference is quite
marked following diastase hydrolysis. This indicates a much greater proportion of starch formed during heterotrophy then during
autotrophy.

There are many marine algae classified as sea-weeds which
synthesize many polysaccharides of commercial interest e.g. agar.

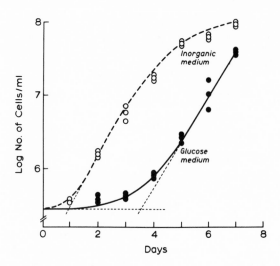

Fig. 7.  The effect of glucose upon the rate of increase of
population density of <u>Chlorella</u> <u>vulgaris</u> (Emerson strain).
(After Griffiths, 1963).

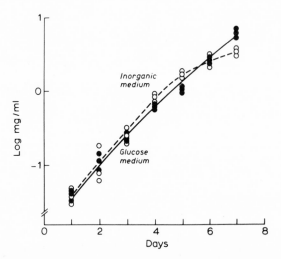

Fig. 8.  The effect of glucose upon the rate of dry weight
accumulation of <u>Chlorella</u> <u>vulgaris</u> (Emerson strain).  Culture
conditions as in Fig. 7.   (After Griffiths, 1963).

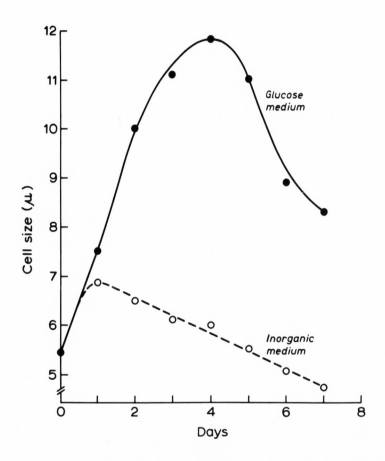

Fig. 9.    The effect of glucose upon cell size of Chlorella
vulgaris (Emerson strain) during growth.  Culture conditions
as in Fig. 7.   (Estimated from the data of Griffiths, 1963).

Table 4. Carbohydrate analysis of autotrophic and heterotrophic cultures of Chlorella vulgaris (Emerson strain) obtained with the starved cells. (After Griffiths, 1965).

| Extraction Procedure | $10^{-9}$ mg Glucose equivalent per cell | | % Glucose equivalent on dry weight basis | |
|---|---|---|---|---|
| | A | H | A | H |
| Alcoholic extract | 0.5 | 5.4 | 0.45 | 0.70 |
| Acid hydrolyzed alcoholic extract | 0.8 | 16.3 | 0.71 | 2.12 |
| Aqueous extract after acid hydrolysis of cell residue | 7.6 | 204.1 | 6.77 | 26.49 |
| Aqueous extract after diastase hydrolysis of cell residue | 2.7 | 123.7 | 2.41 | 16.05 |

A: autotrophic growth
H: heterotrophic growth

These polysaccharides are reviewed by D.F. Jackson (1964) in Algae and Man, a Plenum Press publication.

## Protein Production

Under specialized growth conditions protein content of algal cells of 88% can be attained (Milner, 1953). In heterotrophically grown algae, Samejima et al (1958) obtained 49% protein for Chlorella pyrenoidosa grown on glucose. These protein levels are only obtained in the presence of excess sources of inorganic nitrogen.

Of practical interest, the work of Griffiths (1967) on protein production by heterotrophic culture of algae is presented as follows. Emerson strain of Chlorella vulgaris was grown in a medium which consists of 0.5% peptone and an otherwise nitrogen-free inorganic medium supplemented by 0.05M glucose. In testing inorganic nitrogen during heterotrophic culture, potassium nitrate 0.025M was substituted for peptone. Cells were first kept in dark on inorganic medium for starvation, then inoculated with a relatively high algal population. The results are shown in Table 5. Peptone is far more effective than nitrate as a source of nitrogen. Peptone allows the production of a greatly increased cellular material as measured in dry weight or in population density.

The amounts of soluble and the insoluble nitrogen synthesized under autotrophic and heterotrophic conditions were compared. The results are shown in Table 6.

By assuming that the insoluble nitrogen is derived largely from cellular protein, it is clear that peptone-supplied heterotrophic culture gives a significantly higher content of protein. Compared with the autotrophic cultures, the soluble nitrogen value in peptone-supplied systems is only slightly less.

## Nitrogen Fixation

Increased field fertility through nitrogen fixation is attributed mainly to species of blue-green algae, Azotobacter and symbiotic nitrogen fixing bacteria. The effect of inoculating paddy fields of rice with nitrogen fixing alga Tolypothrix tenuis was studied by Watanabe (1962). The average increase of rice crop was 2.0% the first year, and 8.0, 15.1, 19.5 and 10.5% for each of the subsequent years.

Certain nitrogen fixing algae, such as Nostoc muscorum (Allison et al, 1937), Tolypothrix tenuis (Kiyohara et al, 1960),

Table 5.  Heterotrophic culture of *Chlorella vulgaris* (Emerson strain) on nitrate and peptone.  (After Griffiths, 1967).

| Algal Cells | Nitrogen Source | |
|---|---|---|
| | Nitrate | Peptone |
| Mean population density ($10^6$/ml) | 0.266 | 3.136 |
| Algal volume ($10^6 \mu^3$/ml) | 219 | 1522 |
| Mean cell volume ($\mu^3$) | 823 | 485 |
| Algal dry weight ($10^{-3}$ mg/ml) | 169 | 811 |
| Mean cell dry weight ($10^{-7}$ mg) | 6.4 | 2.6 |

Table 6. Nitrogen content of autotrophic and heterotrophic grown cells of _Chlorella vulgaris_ (Emerson strain) (After Griffiths, 1967).

| Trophic Type | Nitrogen Content (% dry weight) | |
| --- | --- | --- |
| | Soluble | Insoluble |
| Autotrophy | 1.2 | 7.4 |
| Heterotrophy | | |
| nitrate | 0.2 | 1.3 |
| peptone | 0.9 | 2.6 |

Chlorogloea fritschii (Fay, 1965), and Anabaenopsis circularis
(Watanabe et al, 1967) are capable of heterotrophic growth.  However, the maximum values for the growth rate and final growth yield
obtained were far less than those obtained in the usual autotrophic
culture (Watanabe et al, 1967).  For theoretical reasons experiments
on heterotrophic nitrogen fixation by the blue-green algae were
completed by Fay (1965) for Chlorogloea fritschii and by Watanabe
et al (1967) for Anabaenopsis circularis.

Fay observed Chlorogloea fritschii fixed elemental nitrogen
to a limited extent in dark in a strictly inorganic medium.  Growth
as well as nitrogen fixation continued in the dark when a suitable
organic substrate was added.  Among the organic substrates tested,
sucrose was the most readily utilized and was outstanding in supporting nitrogen fixation in dark.  In light, sucrose assimilation
proceeded more vigorously and resulted in a fourfold increase in
the rate of growth and nitrogen fixation.  Sucrose assimilation
was increased in light in the absence of carbon dioxide in the gas
phase.  Nitrogen fixation was greatest when the alga was supplied
with sucrose and carbon dioxide.

Another nitrogen fixing alga Anabaenopsis circularis, which
was observed by Watanabe et al (1967) grows and fixes nitrogen at
the greatest rate on glucose in the absence of organic nitrogen.
As organic carbon sources, fructose, sucrose and maltose were also
able to support growth and nitrogen fixation, but all were inferior
to glucose.  The effect of light on the growth was examined with
fructose and was found to be beneficial.


                                  SUMMARY

Data presented herein only include algae which grow on at
least one organic substrate which may be used as either a source
of carbon or nitrogen.  There are several different genera, species
and even strains of algae which grow on organic substrates.  Organic substrates used are quite specific.  Monosaccharides and acetate
are more widely utilized than any other organic substrates.  Although organic nitrogen sources are not always necessary, they are
more effective in many cases than inorganic nitrogen, e.g. casamino
acids in Tolypothrix tenuis (Kiyohara et al, 1960), peptone in
Chlorella vulgaris (Griffiths, 1967).  There are certain algae
which have an absolute requirement for organic nitrogen e.g. yeast
autolyzate in Prototheca zopfii (Barker, 1935).

The organic substrates, whether simple or complex are assimilated and utilized for conversion to energy and into cellular material.
If the objective is to produce cellular material itself, then the
autotrophic or heterotrophic process, which gives the highest
efficiency of conversion into cellular material should be used.

With respect to substrate assimilation, a mutual quantitative or qualitative relation exists between the carbon and nitrogen source. For example, the amount of available nitrogen has an enhancing effect on the assimilation of sugar in Chlorococcum macrostigmatum (Maloney, 1959). In order to provide a more practical basis for the heterotrophic culture of algae, nutritional patterns as well as suitable growth conditions have been summarized (Table 1).

Studies on heterotrophic nutrition have also been extended to those substrates in organic wastes such as spent sulfite liquor (Maloney, 1959), while other workers have been investigating heterotrophic growth of algae in waste-stabilization ponds (Eppley et al, 1962, 1963a, 1963b; Wiedeman et al, 1965). Since economy is the most important consideration in any application, selection of algae and a process should be based upon maximum productivity and the utilization of cheap substrates. The cheapest organic substrates are waste materials in the industrial wastes or sewage. Here algae can be used for both waste purification and protein synthesis.

If one emphasizes organic sources of carbon or nitrogen, purified glucose or peptone are preferred. These are relatively expensive, therefore, further studies seeking suitable substrates from waste liquors are required. Product oriented processes e.g. xanthophylls, carotene, polysaccharides etc. should also result in economies.

Furthermore, for more economical utilization of organic substrates, light should be introduced since in almost every instance tested, it stimulated growth.

In comparison to autotrophic culture, heterotrophic conditions gave higher yield of particular products, e.g. starch in the Emerson strain of Chlorella vulgaris (Griffiths, 1965). As emphasized by Griffiths (1963), heterotrophically produced algal cells yield more polysaccharides rather than other cellular constituents. Considerable research is needed to obtain information on other polysaccharides synthesized by uni-cellular algae.

Algae, characterized by the usually high protein content of about half the biomass, may have potential use as food. In peptone-grown culture, the percentage protein on dry weight basis is only slightly below that in the autotrophic cultures. Between nitrate and peptone, the latter gives four times as much soluble nitrogen and twice as much insoluble nitrogen (Griffiths, 1967). Nutritional studies must be completed evaluating feeding efficiency and toxicity of algal proteins.

In regard to food production for higher plants, another important function of algae is nitrogen fixation. The nitrogen

fixing algae can by-pass the costly requirement for addition of
inorganic or organic sources of nitrogen to soil for increasing
productivity. The fact that certain nitrogen fixing algae can be
grown heterotrophically in dark may be of practical interest
(Watanabe et al, 1967), where large quantities of algal materials
are produced for agricultural purposes in fermentation tanks.

From an engineering standpoint, the larger the cell the less
the power needed to concentrate the biomass during harvesting.
Therefore a larger cell size may be preferable. A systematical
study on the type of nutrition as a factor of cell size has been
studied by Andreeva (1967) for Chlorella vulgaris. Cells of some
strains were larger in heterotrophic and mixotrophic cultures than
in autotrophic culture. It was shown that the appearance of giant
cells in Chlorella is associated not only with the nature of strain
itself, but also with nutritive conditions. As indicated in
Figure 9 for starch accumulation by the Emerson strain of Chlorella
vulgaris, the average cell size in heterotrophic is much larger
than those obtained in autotrophic culture. This feature would
provide an additional advantage.

In conclusion, with an expanding industry and world population,
the use of organic wastes as substrates for algal culture for supply
of protein and products is necessary. Details are needed in developing product oriented processes in which mixotrophic conditions of
growth are utilized. Additional process development and bioengin-
eering are needed in these areas.

## REFERENCES

Algeus, S., 1948a. Glycocoll as a source of nitrogen for Scenedesmus
    obliquus. Physiol. Plant. 1: 65-84.
Algeus, S., 1948b. The utilization of glycocoll by Chlorella
    vulgaris. Physiol. Plant. 1: 236-244.
Algeus, S., 1948c. The deamination of glycocoll by green algae.
    Physiol. Plant. 1: 382-386.
Algeus, S., 1949. Alanine as a source of nitrogen for green algae.
    Physiol. Plant. 2: 266-271.
Allison, F.E., Hoover, S.R. and Morris, H.J., 1937. Physiological
    studies with the nitrogen-fixing alga, Nostoc muscorum. Bot.
    Gaz. 98: 433-463.
Allison, R.K., Skipper, H.E., Reid, M.R., Short, W.A. and Hogan,
    G.L., 1953. Studies on the photosynthetic reaction. I. The
    assimilation of acetate by Nostoc muscorum. J. Biol. Chem.
    204: 197-205.
Anderson, E.H., 1945. Studies on the metabolism of the colorless
    alga Prototheca zopfii. J. Gen. Physiol. 28: 297-327.

Andreeva, V.M., 1967.  Variability of taxonomic characters of
     unicellular green algae in culture. II. Type of nutrition as
     a factor of cell size in Chlorella vulgaris.  Bot. Zh. 52:
     960-966.
Bailey, J.M. and Neisch, A.C., 1954.  Starch synthesis in Chlorella
     vulgaris.  Can. J. Biochem. Physiol. 32:  452-464.
Barker, H.A., 1935.  The metabolism of the colorless alga, Proto-
     theca zopfii Krüger. J. Cell. Comp. Physiol. 7:  73-79.
Barker, H.A., 1936.  The oxidative metabolism of the colorless
     alga, Prototheca zopfii.  J. Cell. Comp. Physiol. 8:   231-250.
Barker, S.A. and Bourne, E.J., 1955.  Composition and synthesis
     of the starch of Polytomella coeca. pp. 45-56.  In S. H.
     Hunter & A. Lwoff (ed.), Biochemistry and Physiology of
     Protozoa Vol. II. Academic Press, New York.
Beijerinck, M.W., 1898.  Notiz über Pleurococcus vulgaris.
     Centralbl. f. Bakt. II. 4:   785-787.
Belcher, J.H., and Fogg, G.E., 1958.  Studies on the growth of
     Xanthophyceae in pure culture. III. Tribonema aequale Pascher.
     Arch. Mikrobiol. 30:   17-22.
Belcher, J.H. and Miller, J.D.A., 1960.  Studies on the growth of
     Xanthophyceae in pure culture. IV. Nutritional types amongest
     the Xanthophyceae.  Arch. Mikrobiol. 36:   219-228.
Bergmann, L., 1955.  Stoffwechsel und mineralsalzernährung
     einzelliger Grundlagen. II. Vergleichende Untersuchen über
     den Einfluss mineralischer Faktoren bei heterotropher und
     mixotropher Ernährung. Flora (Jena) 142:   493-539.
Bishop, N.I., 1961.  The photometabolism of glucose by an hydro-
     gen-adapted alga.  Biochem. Biophys. Acta. 51:   323-332.
Blum, J.J., Podolsky, B. and Hutchens, J.O., 1951.  Heat production
     in Chilomonas.  J. Cell. Comp. Physiol. 37:   403-426.
Bristol-Roach, B.M., 1928.  On the influence of light and of
     glucose on the growth of a soil alga.  Ann. Bot. 42:   317-345.
Claes, H., 1957.  Biosynthese von Carotinoiden bei Chlorella. III.
     Untersuchungen über die lichtabhängige Synthese von α-und
     β-Carotin und Xanthophyllen bei der Chlorella-Mutante 5/520.
     Z. Naturforsch. 12b:   401-407.
Claes, H., 1959.  Biosynthese von Carotinoiden bei Chlorella. V.
     Die Trennung von Licht- und Dunkelreaktionen bei der
     lichtabhängigen Xanthophyllsynthese von Chlorella. Z.
     Naturforsch. 14b:   4-7.
Clemencon, H., 1965.  Kultur und Entwicklung von Chlorocloster
     engadinensis Vischer (Xanthophyceae).  Arch. Mikrobiol. 51:
     199-212.
Cosgrove, W.B., 1950.  Studies on the question of chemoautotrophy
     in Chilomonas paramecium. Physiol. Zoöl. 23:  73-84.
Cosgrove, W.B. and Swanson, B.K., 1952.  Growth of Chilomonas
     paramecium in simple organic media. Physiol. Zoöl. 25:
     287-292.

Cramer, M. and Myers, J., 1952. Growth and photosynthetic charac-
    teristics of Euglena gracilis. Arch. Mikrobiol. 17: 384-402.
Dvorakova-Hladka, J., 1966. Utilization of organic substrates
    during mixotrophic and heterotrophic cultivation of algae.
    Biol. Plant. (praha) 8: 354-361.
Emerson, R., 1927. The effect of certain respiratory inhibitors
    on the respiration of Chlorella. J. Gen. Physiol. 10:
    469-477.
Eppley, R.W. and MaciasR, F.M., 1962. Rapid Growth of sewage
    lagoon Chlamydomonas with acetate. Physiol. Plant. 15:
    72-79.
Eppley, R.W., Gee, R. and Saltman, P., 1963a. Photometabolism of
    acetate by Chlamydomonas mundana. Physiol. Plant 16: 772-
    792.
Eppley, R.W. and MaciasR, F.M., 1963b. Role of the alga Chlamydo-
    monas mundana in anaerobic waste stabilization lagoons.
    Limnol. Oceanogr. 8: 411-416.
Farrow, W. and Tabenkin, B., 1966. Process for the preparation of
    lutein. U.S. Patent 3,280,502.
Fay, P., 1965. Heterotrophy and nitrogen fixation in Chlorogloea
    fritschii. J. Gen. Microbiol. 39: 11-20.
Fay, P. and Fogg, G.E., 1962. Studies on nitrogen fixation by
    blue-green algae. III. Growth and nitrogen fixation in
    Chlorogloea fritschii Mitra. Arch. Mikrobiol. 42: 310-321
Finkle, B.J., Appleman, D. and Fleischer, F.K., 1950. Growth of
    Chlorella vulgaris in the dark. Science 111: 309.
Fogg, G.E., 1956. The comparative physiology and biochemistry of
    the blue-green algae. Bact. Rev. 20: 148-165.
Frederick, J.F., 1951. Preliminary studies on the synthesis of
    polysaccharides in the algae. Physiol. Plant. 4: 621-626.
Fujita, K., 1959. The metabolism of acetate in Chlorella cells.
    J. Biochem. (Tokyo) 46: 253-268.
Gaffron, H. and Rubin, J., 1942. Fermentative and photochemical
    production of hydrogen in algae. J. Gen. Physiol. 26:
    219-240.
Griffiths, D.J., 1960. The heterotrophic nutrition of Chlorella
    vulgaris (Brannon No. 1 strain). Ann. Bot. N.S. 24: 1-11.
Griffiths, D.J., 1963. The effect of glucose on cell division
    in Chlorella vulgaris, Beijerinck (Emerson strain). Ann.
    Bot. N.S. 27: 493-504.
Griffiths, D.J., 1965. The accumulation of carbohydrate in
    Chlorella vulgaris.under heterotrophic conditions. Ann. Bot.
    N.S. 29: 347-357.
Griffiths, D.J., 1967. The effect of peptone on the growth of
    heterotrophic cultures of Chlorella vulgaris (Emerson strain)
    Planta (Berl.) 75: 161-163.
Guehler, P.F., 1966. Transformation of steroids by algae.
    Dissertation Absts. 26: 6371-6372.
Guehler, P.F., Dodson, R.M. and Tsuchiya, H.M., 1962. Transforma-
    tion of steroids with Chlorella pyrenoidosa. Microbiol. 48:
    377-379.

Harder, R., 1917. Ernährungsphysiologische Untersuchungen an
    Cyanophyceen, hauptsächlich dem endophytischen Nostoc
    punctiforme. Z. Bot. 9: 145-242.
Holton, R.W., 1968. Fatty acids in blue-green algae: Possible
    relation to phylogenetic position. Science 160: 545-547.
Hutchens, J.O., 1940. The need of Chilomonas paramecium for iron.
    J. Cell. Comp. Physiol. 16: 265-267.
Hutchens, J.O., 1948. Growth of Chilomonas paramecium in mass
    cultures. J. Cell. Comp. Physiol. 32: 105-116.
Hutchens, J.O., Podolsky, B. and Morales, M.F., 1948. Studies on
    the kinetics and energetics of carbon and nitrogen metabolism
    of Chilomonas paramecium. J. Cell. Comp. Physiol. 32: 117-
    141.
Iggena, M.L., 1938. Beobachtungen über die Wirkung des Lichtes
    auf das Wachstum von Blaualgen und Grünalgen. Arch. Mikrobiol.
    9: 129-166.
Karlander, E.P. and Krauss, R.W., 1966. Responses of heterotrophic
    cultures of Chlorella vulgaris Beijerinck to darkness and
    light. II. Action spectrum for and mechanism of the light
    requirement for heterotrophic growth. Plant. Physiol. 41:
    7-14.
Kathrein, H.R., 1960. Production of carotenoids by the cultivation
    of algae. U.S. Patent 2,949,700.
Kathrein, H.R., 1963. Production of carotenoid pigments. U.S.
    Patent 3,108,402.
Kathrein, H.R., 1964. Production of carotenoids by the cultivation
    of algae. U.S. Patent 3,142,135.
Killam, A. and Myers, J., 1956. A special effect of light on the
    growth of Chlorella vulgaris. Amer. J. Bot. 43: 569-572.
Kiyohara, T., Fujita, Y., Hattori, A. and Watanabe, A., 1960.
    Heterotrophic culture of a blue-green alga, Tolypothrix tenuis
    I. J. Gen. Appl. Microbiol. 6: 176-182.
Kratz, W.A. and Myers, J., 1955. Photosynthesis and respiration
    of three blue-green algae. Pl. Physiol. Lancaster 30: 275-
    280.
Lewin, J.C., 1953. Heterotrophy in diatoms. J. Gen. Microbiol.
    9: 305-313.
Lewin, R.A., 1954. The utilization of acetate by wild-type and
    mutant Chlamydomonas dysosmos. J. Gen. Microbiol. 11: 459-
    471.
Lewin, J.C. and Lewin, R.A., 1960. Auxotrophy and heterotrophy
    in marine littoral diatoms. Can. J. Microbiol. 6: 127-134.
Luedemann, G., Charney, W., Woyciesjes, A., Pettersen, E.,
    Peckham, W.D., Gentles, M.J., Marshall, H. and Herzog, H.L.,
    1961. Microbiological transformation of steroids. IX.
    The transformation of Reichstein's Compound S by Scenedesmus
    sp.: Diketopiperazine metabolites from Scenedesmus sp. J.
    Org. Chem. 26: 4128-4129.
Lwoff, A., 1947. Some aspects of the problem of growth factors for
    protozoa. Ann. Rev. Microbiol. 1: 101-114.

Lynch, V.H. and Calvin, M., 1953.  $CO_2$ fixation by Euglena.  Ann. N.Y. Acad. Sci. 56:  890-900.

Maclachlan, G.A. and Porter, H.K., 1959.  Replacement of oxidation by light as the energy source for glucose metabolism in tobacco leaf. Proc. Roy. Soc. B 150:  460-473.

Maloney, T.E., 1959.  Utilization of sugars in spent sulfite liquor by a green alga, Chlorococcum macrostigmatum. Sewage Industr. Wastes 31:  1395-1400.

McBride, L.J., 1961.  The stimulatory effect of low intensity light on the heterotrophic growth of Chlorella.  Dissertation Absts. 22:  1805-1806.

Milner, H.W., 1953.  The chemical composition of algae.  Ch. 19. In J.S. Burlew (ed.), Algal culture from laboratory to pilot plant.  Carnegie Inst. Wash. Pub. 600.  Washington, D.C.

Mineeva, L.A., 1962a.  Effect of pH on the autotrophic and heterotrophic nutrition of Chlorella vulgaris and Scenedesmus obliquus.  Mikrobiologiya 31:  233-240.

Mineeva, L.A., 1962b.  The effect of light intensity upon auto- trophic and heterotrophic nutrition of Chlorella vulgaris and Scenedesmus obliquus.  Mikrobiologiya 31:  411-416.

Myers, J., 1947.  Oxidative assimilation in relation to photo- synthesis in Chlorella. J. Gen. Physiol. 30:  217-227.

Myers, J. and Johnston, J.A., 1949.  Carbon and nitrogen balance of Chlorella during growth.  Plant Physiol. 24:  111-119.

Nakamura, H., 1961.  Report on the present situation of the Microalgae Research Institute of Japan.  The Japan Nutrition Association.  Tokyo. 2:  1-12.

Neish, A.C., 1951.  Carbohydrate nutrition of Chlorella vulgaris. Can. J. Bot. 29:  68-78.

Oswald, W.J., and Gotaas, H.B., 1957.  Photosynthesis in Sewage Treatment.  Tr. Amer. Soc. Civil Eng. 122:  73-97.

Pace, D.M., 1948.  The utilization of ethyl alcohol as source of carbon in Chilomonas paramecium.  Anat. Rec. 101:  730-731.

Parker, B.C., Bold, H.C. and Deason, T.R., 1961.  Facultative heterotrophy in some chlorococcacean algae.  Science 133: 761-763.

Pearce, J. and Carr, N.G., 1969.  The incorporation and metabolism of glucose by Anabaena variabilis.  J. Gen. Microbiol. 54: 451-462.

Pearsall, W.H. and Bengry, R.P., 1940.  Growth of Chlorella in relation to light intensity.  Ann. Bot. N.S. 4:  485-494.

Pringsheim, E.G., 1955.  Kleine Mitteilungen über Flagellaten und Algen. II. Euglena gracilis var. saccharophila n. var. und eine vereinfachte Nährlösung zur Vitamin $B_{12}$ Bestimmung. Arch. Mikrobiol. 21:  414-419.

Pringsheim, E.G. and Wiessner, W., 1960.  Photoassimilation of acetate by green organisms.  Nature 188:  919-921.

Pruess, L., Arnow, P., Wolcott, L., Bohonos, N., Oleson, J.J. and Williams, J.H., 1954.  Studies on the mass culture of various algae in carboys and deep-tank fermentations. Appl. Microbiol. 2:  125-130.

Reazin, G.H., 1956.  The metabolism of glucose by the algae.
    Ochromonas malhamensis.  Plant Physiol. 31:  299-303.
Samejima, H. and Myers, J., 1958.  On the heterotrophic growth of
    Chlorella pyrenoidosa.  J. Gen. Microbiol. 18:  107-117.
Schlegel, H.G., 1956.  Die Verwertung von Essigsäure durch Chlorella
    im Licht. Z. Naturforsch. 14b:  246-253.
Taylor, F.J., 1950.  Oxidative assimilation of glucose by
    Scenedesmus quadricauda. J. Exptl. Bot. 1:  301-321.
Theriault, R.J., 1965.  Heterotrophic growth and production of
    xanthophylls by Chlorella pyrenoidosa.  Appl. Microbiol.
    13:  402-416.
Watanabe, A., 1962.  Effect of nitrogen fixing blue-green alga,
    Tolypothrix tenuis on the nitrogenous fertility of paddy soil
    and on the crop yield of rice plant.  J. Gen. Appl. Microbiol.
    8:  85-91.
Watanabe, A. and Yamamoto, Y., 1967.  Heterotrophic nitrogen fixa-
    tion by the blue-green alga Anabaenopsis circularis. Nature
    214:  738.
Whatley, F.R. and Losada, M., 1964.  The photochemical reactions
    of photosynthesis. Chap. 5.  In. A.C. Giese (ed.), Photo-
    physiology Vol. 1. Academic Press, New York.
Wiedeman, V.E. and Bold, H.C., 1965.  Heterotrophic growth of
    selected waste-stabilization pond algae.  J. Phycol. 1:
    66-69.
Wiessner, W. and Gaffron, H., 1964.  Role of photosynthesis in the
    light-induced assimilation of acetate by Chlamydobotrys.
    Nature 201:  725-726.
Wilson, B.W. and Danforth, W.F., 1958.  The extent of acetate and
    ethanol oxidation by Euglena gracilis. J. Gen. Microbiol.
    18:  535-542.
Zajic, J.E., 1969.  Xanthophylls formation in heterotrophically
    grown algae. Unpublished.  8 pp.

ALGAL PRODUCTS

B. Volesky, J.E. Zajic,and E. Knettig

Biochemical Engineering, University of Western Ontario

London, Ontario

## INTRODUCTION

Algae are sources of many useful products.  Many polysaccharides are recovered from algae (Percival and McDowel, 1967; Fogg, 1966).  The most important of these are alginic acid (Haug and Larsen, 1964; Peat and Turvey, 1965; Kim et al., 1965), laminarin, fucoidin, galactans, agar (Percival, 1964; de Leon et al., 1963; Ming, 1962; Painter, 1960; Fogg, 1953), carrageenin, xylans and mannans.  Starch is also produced but not in large quantities. Marine algae are excellent sources of potash, iodine and trace elements.  Algae are used for food and some cultures are very high in protein, vitamins and essential growth substances.  Many pigments are produced, but only carotenoids having pro-vitamin-A activity are of current interest.  Algae are used as fodder and fertilizer.  Some produce medicinal products, although no antibiotics are being produced commercially from algae.  Diatomaceous deposits are used to recover high yields of silica.  These materials are only a few of the more important products produced by algae.

## Algal Polysaccharides

Extensive investigations have been taking place during the past two decades in the field of algal polysaccharides.  Major marine plant polysaccharides are mucilaginous in nature.  A wide spectrum of these complex polymers have been isolated from both marine and fresh water algae.  Mucilage extruded by marine algae which is dried or exposed to fresh water has a very high content of polysaccharide.  Polysaccharides when extracted from algal tissue may make up to 10-65% of the dry weight, depending on the

genera and the conditions of growth.  Environmental factors are
very active in controlling polymer synthesis.

The algal polysaccharides of greatest industrial importance
from brown algae are (Schmid, 1962); alginic acid derivatives,
fucoidans, funoran, laminarans, and from red algae: sulphated galactans agar and carrageenin.  Other polymers such as xylans, mannans, xylomannan and glucan, (Ralph and Bender, 1965), eucheuman,
furcellaran, hypean, iridophycan, phyllophoran, odonthalan (Usov
and Kochetkov, 1968), cellulose, chitin, hemicelluloses, pectins
and mucilages are primarily of academic interest.  Most are components of the cell wall.  Algal starch has been known as a storage product in many genera (Gusev, 1961; Love et al., 1963;
Meeuse et al., 1960).

Many structurally similar polysaccharides isolated from separate genera differ only slightly from known polymers.  Their
structural composition and uses are still being extensively
studied.


                          Alginates

Discovery of these substances is attributed to E.C. Stanford
in 1883, who found that many of the brown seaweeds contained a
viscous matter which on treatment with sodium carbonate and subsequently with mineral acid, produced a new compound which was
called alginic acid.  This is a reserve product of cellular metabolism, (Doshi and Rao, 1968).  The production of this compound
and its salts has become the major product of the kelp industry.

Alginic acid is a complex polyuronide which is very stable
to hydrolysis.  It has been shown to be comprised mainly of $\beta$-1,4'-
linked D-mannuronic and L-guluronic acid units as shown in Figure
1 (Percival, 1964).  The fine details of structure of the macro-
molecule remain to be elucidated.

The alginic acid itself is insoluble in water but has a high
capacity for absorbing water.  Its Na, K, $NH_4$, Mg, Fe(ferrous)
salts are soluble in water, giving high viscosities, whereas the
Ca, Al, Zn, Cu, Cr, Fe(ferric) and Ag complexes yield insoluble
compounds.

The alginates of the heavy metals form plastic type materials
when wet and set up hard on drying.  Alginic acid derivatives are
extracted from Phaeophyta (Rae and Mody, 1968) of different genera
according to local conditions and occurence.  Brown algae harvested for alginates in various parts of the world are as follows:

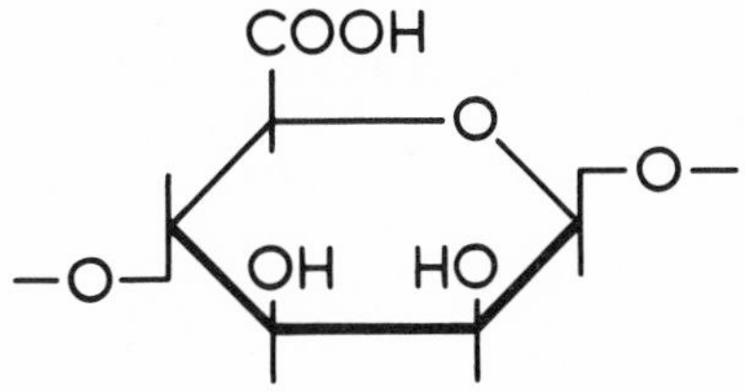

$\beta$-1,4-D-

MANNURONIC ACID

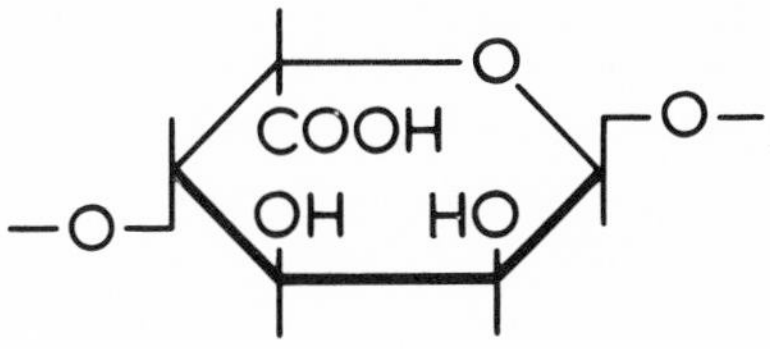

$\beta$-1,4-L-

GULURONIC ACID

Fig. 1.  Structures of mannuronic and guluronic acids.

| | |
|---|---|
| Laminaria, Ascophyllum | – Northern coasts (Haug, 1965) |
| Macrocystis | – California coasts |
| Cystoseira barbata | – Black Sea (Yatsenko, 1962; Tsapko, 1966) |
| Eklonia, Eisenia | – Japan coasts |
| Macrocystis, Eklonia | – Australasia coasts |
| Macrocystis, Lessonia, Durvillea | – S. America coasts |
| Laminaria pallida, Eklonia | – S. Africa coasts |

The amount of alginic acid in the various Fucales and laminarians varies with the season.  It is lowest in spring and highest in the autumn.  It is also affected by habitat (i.e. exposure, protection, currents, and depth).  In Macrocystis there is evidence of a variation with latitude.  This is apparently associated with temperature and light variations, since alginates are regarded as one of the products of metabolism.  Positionally there is almost no difference in location of the alginates within the algal body (Miyake, 1959) and it has been isolated from stipe and at least 3 different parts of the frond in approximately equal amounts.  This observation has certain significance from the viewpoint of harvesting and further extraction.  X-ray microradiographs and spodograms have shown that alginic acid is located in all walls (Frei and Preston, 1962).

The harvesting of the seaweed is carried out according to the general occurrence and habitat.  Laminaria spp. as well as Ascophyllum are the most difficult to harvest because they attach to rock surfaces under water.  For this reason they are often

harvested along the shore line from materials cast up during
storms.  Macrocystis which is the main source of alginates on
California coast grows attached by huge holdfasts.  Populations
of this alga reproduces itself vegetatively rather than sexually.
Conditions for sexual reproduction are not favorable due to the
shade under the dense canopy of fronds and leaves floating on the
surface of water.  Growth rate of this particular alga has been
reported to be 10-15 feet/week which is greater than for any land
or sea plant.

Floating mechanical harvesters are employed for harvesting.
The fronds that are cut regenerate and the bed of algal vegetation
may be maintained in growing state for long periods of time.  The
mucilaginous exudate found on the surface of brown seaweeds exposed to the atmosphere comprises high amount of alginic acid
(Kroes, 1968).  Various methods of extraction of seaweed have been
devised.  After desalination and shredding it is leached first
with acid and then treated with soda ash or sodium carbonate solution yielding a crude mass of either raw alginic acid or more
usually calcium or sodium alginates.  The various forms can be
interconverted after priliminary filtering and bleaching.  The
most commonly employed method is that of Green (cold process)
yielding sodium or other alginic salts and that of Gloahec-Herter
yielding alginic acid after removing of laminarin, mannitol and
other salts by treatment with calcium chloride (Figure 2).

Variety of properties of different salts of alginic acid gives
them a remarkably wide range of application for industry use.
Textile industry has been experimenting with seaweed rayons.  Viscous solution of the alkali salts can be spun into artificial silk
threads.  The final yarn is usually of calcium or sodium alginate.
The "disappearing fibre" technique has been made possible because
of the property of the yarn which enable it to dissolve in soap
and soda.  Very fine wools may be strengthened by the addition of
soluble alginate thread which can afterwards be easily washed out
(textile sizing).  Resistant threads can be obtained by replacing
the calcium or sodium with chromium or beryllium.  In practice it
is found easier to weave the yarn in the form of calcium alginate
and then convert the woven material to beryllium or chromium salt
by immersing it in a bath of beryllium or chromium acetate.  An
interesting feature of alginate fabrics is that they are non-
flammable.

As plastic materials, alginates have a number of applications
in the field of medicine.  Particularly accurate dental impressions
may be taken by using alginates.  Gauze made of alginic acid salts
is used which is haemostatic and slowly absorbed in the body.

Because of its non-toxicity and colloidal properties alginic
acid and its salts find considerable use in the food industry.

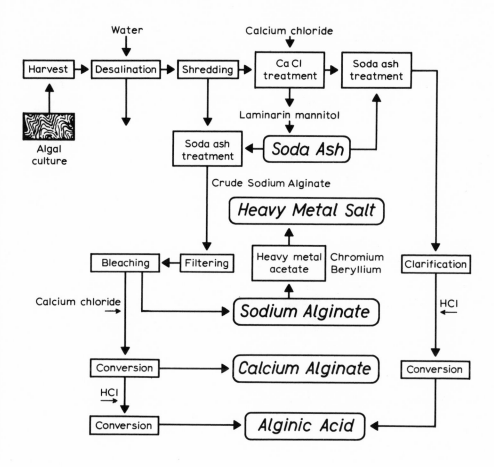

Fig. 2   Schematic for alginic acid production.

By proper control it is possible to produce gels of different viscosities.  Sodium alginate is considered to be superior to all other colloids as a stabilizer and creaming agent for ice cream.

Its thickening properties make it useful in the preparation of foods such as soups, sauces and creams of various kinds.  Films of alginates may also be made into sausage casings.  It is also used as a gel in the freezing of fish.  Antibiotics (aureomycin) may be incorporated into algin gels to keep dispersed.

Alginates are used in pharmaceutical emmulsions and as a
desintegrant in the coating of pills. They swell readily in a
watery medium. The cosmetic trade uses alginates to thicken
various creams. Other applications are as (1) surface films in
the paper industry and (2) in placing a glaze on ceramics. In
emulsification it finds use in latex and other emmulsion paints
as well as a polishing material.

More recently, experiments have been conducted to ascertain
its usefulness as an ameliorator of soils low in organics. In
this respect it competes with materials such as krillium which
improves the physical state of the soil.


### Laminarin

Laminarin was first reported by Schmiedeberg in 1885. It
is neutral polysaccharide which readily hydrolyses. It is composed mainly of $\beta$-1,3'-linked glucose units and was originally
believed to have the structure shown in Figure 3. Some evidence
has been shown that it also contains a small proportion of mannitol and of mannose, as well as 1,6'-linkages. Laminarin is a
mixture of two different molecules, one terminated by a reducing
glucose unit and the other by a non-reducing sugar alcohol such
as mannitol (Dillon, 1964).

Goldstein et al. (1959) suggested that the two molecules
have a non-reducing and reducing arrangements as shown (Percival,
1964):

Non-reducing laminarin:     $\begin{pmatrix} G = D\text{-glucose} \\ M = \text{mannitol} \end{pmatrix}$

$$G1 \xrightarrow{\ \beta\ } [3G1]_n \xrightarrow{\ \beta\ } 6G1 \xrightarrow{\ \beta\ } [3G1]_m$$

$$\begin{array}{c} 1 \\ M \\ 2 \end{array}$$

$$G1 \xrightarrow{\ \beta\ } [3G1]_n \xrightarrow{\ \beta\ } 6G1 \xrightarrow{\ \beta\ } [3G1]_m$$

Reducing laminarin:

$$G1 \xrightarrow{\ \beta\ } [3G1]_m \xrightarrow{\ \beta\ } 6G1 \xrightarrow{\ \beta\ } [3G1]_m \xrightarrow{\ \beta\ } 3G$$

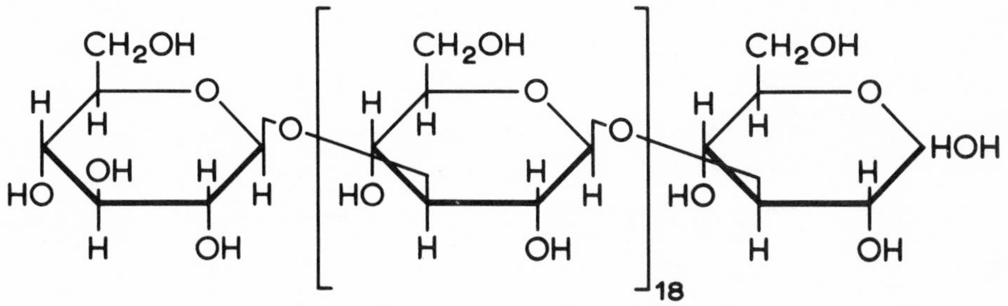

Fig. 3.   General chemical structure of laminarin.

Other analyses indicate that laminarin has a branched rather than a straight chain structure.

There have been isolated "insoluble" and "soluble" forms of laminarin.   The latter was suggested to contain a substance which interferes with the aggregation of the colloidal molecules (Smith and Montgomery, 1959).

Laminarin is chiefly a soluble reserve product occuring mainly in Phaeophyceae e.g. Ascophyllum, Fucus, and Laminaria spp. to the extent of up to 30% of dry weight of the algal tissue (Dillon, 1964).   Oddly, it is not present in some of the members of Phaeophyceae at all.   Evidence of small quantities of the laminarin-type polysaccharide has been presented for the Chloro-phyceae for Ulva lactuca, Acrosiphonia centralis and Cladophora vupestris (Mackie and Percival, 1959).   From a mixed culture of diatoms belonging to the Chrysophyceae crystalline polysaccharide resembling reducing molecule of laminarin has been prepared (Percival, 1967).   Laminarin has not been reported in Rhodophyceae.

## Fucoidin

Fucoidin is a mucilaginous material consisting of a family of compounds extracted from Phaeophyceae.   Over 16% has been reported in Phyllophora nervosa (Gryuner and Evmenova, 1961). It is composed mainly of L-fucose ( a methyl pentose)

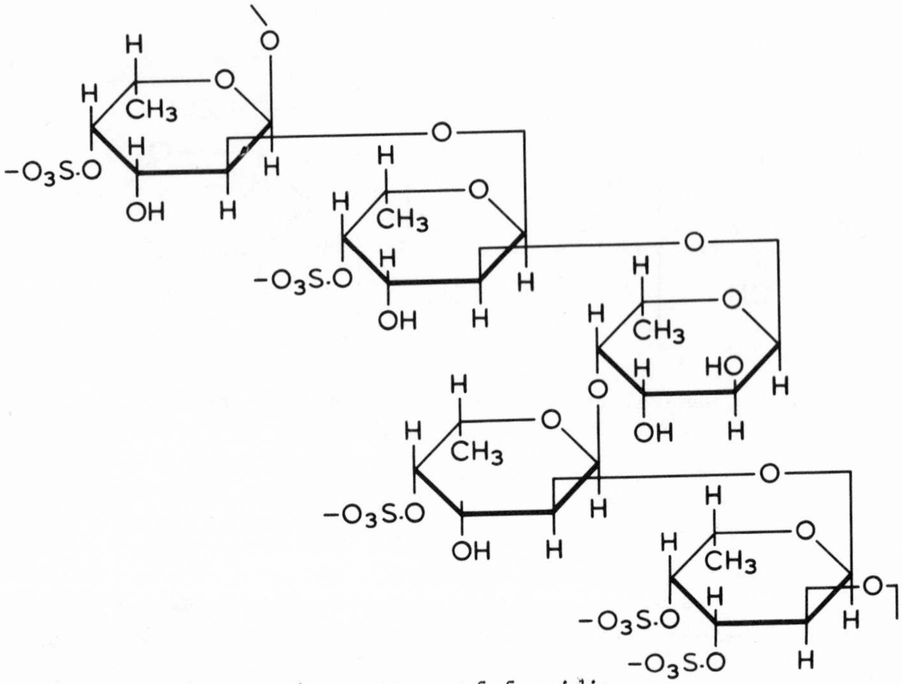

Fig. 4.  Proposed structure of fucoidin.

residues esterified by sulfuric acid.  Carbohydrate units are linked 1,2'- with at least some of the sulphate in $C_4$ position (Figure 2).  Some of the fucose units may be 1,4'-linked (Cote, 1959).

Fucose appears to be the characteristic mucilage sugar of the brown algae.  Species differs in the amounts of mucilage accumulated in a given time and in the relative amounts of the various sugars.

Possible structure of fucoidin is shown in Figure 4, (Fogg, 1953).


### Sulphated Galactans

The red algae synthesize mainly galactan sulphates containing 1,3-linked galactose, together with certain proportions of 3,6-anhydrogalactose residues and sulphate.  The most familiar polysaccharides of the group are agar and carrageenin.

Odonthalan isolated from Pacific <u>Odonthalia corymbifera</u> was found to contain 6-0-methyl-D-galactose (Usov and Kochetkov, 1968). Another highly sulphated methylated galactan called aeodan, has been isolated from <u>Aeodes orbitors</u>. It contains D-galactose, 2-0-methyl-D-galactose and glycerol (Nunn and Parolis, 1968).

Various galactans from different red algae show certain similarities and generally it is possible to conclude that they are a family of polysaccharides based on 1,3-linked galactose and 1,4-linked 3,6-anhydrogalactose. The relative proportions of these residues varies with the species of algae. They also vary as to the proportions of D-, L-sugar and sulfate content.

## Agar

Apart from alginic acid derivatives the other major seaweed industry today is concerned with the production of agar-agar.

Studies on agar (<u>Gelidium amansii</u>) revealed it to be a mixture of 2 polysaccharides agarose and agaropectin. Molecules of agarose are composed of a neutral chain of β-D-galactopyranose residues linked through 1,3-positions, and 3,6-anhydro-α-L-galactopyranose residues connected through 1,4-positions and repeated alternatively.

Agaropectin has the same backbone but also contains ester of sulphate, glucuronic acid and pyruvic acid (Araki, 1965). Polysaccharides like agaropectin, funoran and porphyran resemble agarose structurally in having the disaccharide agarobiose as the structural unit. Specific enzyme agarase has been detected in bacteria which hydrolyzes agar. Dry agar is insoluble in cold and soluble in hot water. Dilute solutions (1-2%) remain liquid down to a temperature between 35-50°C; but gel at lower temperatures. The gel melts at 90°-100°C.

A large number of genera and species of red algae are used as the source of agar and its properties change slightly with the source. The principal source are algae of genus <u>Gelidium</u> (substance extracted is known as gelose), but also some other species of other species of other genera give satisfactorily yields. Some of these algae and where theyooccur are:

| | |
|---|---|
| <u>Gelidium</u> <u>amansii</u> | - Japan, India |
| <u>G. subcostatum</u> | - Japan |
| <u>Ahnfeltia plicata</u> | - Japan, (Arai, 1961) |
| <u>Gelidium</u> <u>cartilagineum</u> | - Pacific coast of America and S. Africa |
| <u>Gelidium lingulatum</u> | - South America |

| | |
|---|---|
| Gracilaria verrucosa | – S. America |
| Gelidium pulchellum | – Ireland |
| Gelidium latifolium | – Ireland |
| Chondrus crispus | – England |
| Gigartina stellata | – England, India |
| Gracilaria confervoides | – Carolina, South Africa, Australia |
| Hypnea musciformis | – Carolina |
| Suluria vittata | – South Africa |
| Pterocladia lucida | – New Zealand |
| P. capillacea | – New Zealand |
| Ahnfeltia | – New Zealand |
| Furcellaria fastigiata | – Baltic Sea (Czapke, 1961) |
| Phyllophora | – Black Sea |
| Eucheuma | – India, Australia |

In some species there is a seasonal periodicity in the amount of gelose the alga contains, e.g. Gelidium cartilagineum in California reaches a peak in June.

The agarophytes can be divided into three groups on the basis of the "setting power of the gel":
1) Gelidium type – decoction sets firm even if dilute.
2) Gracilaria, Hypnea type – decoction sets firm if more concentrated or if electrolytes are added.
3) Chondrus type – only sets firm if very concentrated.

Several approaches to the extraction of agar from the sea-weeds may be taken and industrial extraction procedures differ. Usually extraction is carried out with hot or boiling water. The extract is then treated with active carbon, filtered and the fil-trate allowed to set until it gels. The gel is purified by freez-ing, the water flowing off the impurities during the thawing operation. The pure gel is dried. The yield depends on the source but is generally between 24-45% of the dry weight of the algae (Kappanna and Rao, 1963). Figure 5 describes one of the procedures used in agar production.

Another approach (Steinmetz et al., 1969) is taken by treat-ment of the seaweed below 25°C with alternate alkali and acid washing prior to extraction. Additional purification follows steps described in Figure 5.

It has been reported (Gryuner, 1961) that agaroid extracted from Phyllophora nervosa with boiling water consist of a mixture of carbohydrates, polysaccharides and other organic and mineral compounds. Purified agaroid even contains some nitrogenous mat-erial as well as other impurities. It retains 13-15% mineral constituents.

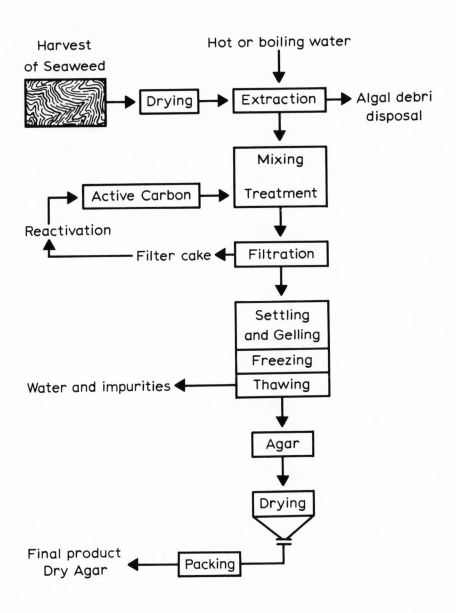

Fig. 5.  Agar production from seaweed.

Washing water divides the agaroid into the true agaroid (~60%)
which is insoluble in cold water dissolves freely in hot water,
and agaroidin, which is soluble in water and 40% alcohol.  Agaroid has a crystalline structure (22.16% sulfate after hydrolysis
and 11.5% sulfate in the ash) while agaroidin is amorphous (15.5%
and 5.56% sulfate) (Gryuner, 1961).

Agar has a variety of uses in commerce and is very important
in microbiological laboratory studies.  A 1-2% solution gives a
satisfactory gel which in the presence of different nutrients
remains undigested and refractile to microbial attack.  Agar is
non-toxic and has gained wide use in the food industry for canning
meat and fish and it is being used in small quantities in some
canned foods and cereals.  It has a property of passing through
the stomach undigested, releasing its water content.  It has
another popular use as an emulsifier in laxatives.  Agar has also
been used in cosmetics, leather, textile industries, etc., in a
manner analagous to alginates and carrageenin.


## Carrageenin

Carrageenin is a cell wall polysaccharide complex extracted
from the Rhodophyceae.  Carrageenin has been fractionated into
two distincts polysaccharides.  (1) κ-carrageenin (kappa) and
(2) λ-carrageenin (lambda).  Both form strong milk gels and mod-
erate water gels.

κ-carrageenin contains sulphated 1,3-linked-β-D-galactose
residues together with 1,4-linked 3,6-anhydro-α-D-galactose residues (Young, 1961; Anderson et al., 1968; Smith and Montgomery,
1959; Clingman and Nunn, 1959).  Evidence has been shown (O'Neil,
1955) of a branching point at $C_6$ for the tenth D-galactose residue to which is attached D-galactose-3,4- or 3,6-disulphate as
a non-reducing terminal residue.  κ-carrageenin does not show
any increase in gel formation after strong or mild alkali hydrolysis.  It is believed to be a polysaccharide of decreased viscos-
ity of lower mol. wt. and less polymerization than its precursor,
λ-carrageenin (Stanley, 1963).

The repeating unit in κ-carrageenin is:

-3 Gal 1 —β— 4-D-A Gal. 2 —α—          S is $S \begin{matrix} O \\ \parallel \\ \parallel \\ O \end{matrix} - OH$

$$Gal = D\text{-}Galactose$$
$$A\ Gal = 3,6\text{-}anhydro\text{-}galactose$$

The λ- carrageenin structure has been less studied and is believed to comprise a highly sulphated 1,3'-linked galactan containing only small proportion of the 3,6-anhydroderivative. The galactans isolated from different algae may differ very slightly but all show similarities to carrageenins (Percival, 1964).

The difference in physical properties between the two carrageenins (one is more soluble in cold water than the other which forms a gel more readily) appears to be due to the content of sodium and calcium (Young, 1961; Schmitt, 1961).

The main source of carrageenins is Chondrus crispus, called sometimes carragheen or "Irish moss", Gigartina stellata, Iridae laminarioides. Galactans of similar composition and properties have been reported also in Dilsea edulis, Porphyra umbicalis, P. capensis, P. crispata, Bangia fuscopurpurea (Percival, 1964).

The procedure of extraction of these types of polysaccharides is very similar to that of agar. Production of carrageenins increases every year. Abundant sources of carrageenin are species of Chondrus and Gigartina located in the intertidal zone in the Maritime Provinces of Canada.

Carrageenin is non-toxic. Its colloidal properties are widely used in many branches of industry to stabilize emulsions and suspend solids, etc. Particular use of carrageenin is therefore found in food, pharmaceutical, brewing, leather and textile industries. Carragheen has a culinary use in the preparation of blancmanges and is used medicinally as a cough tincture.

## Xylans

Xylans are polysaccharides yielding only xylose upon hydrolysis. It has been shown that xylan is one of the essential cell wall constituents of some siphonous green algae (Miwa et al., 1961) as Bryopsis maxima, Caulerpa brachypus, C. racemosa var. lactivireus, Halimeda cuneata, Chorodesmis formosana, Pseudodichotomosiphon constricta, Udotea oricutalis. It comprises mainly β-1,3'-linked residues (Percival, 1964).

$$X1 \xrightarrow{\beta} 3 \; X \; 1 \xrightarrow{\beta} 3 \; X \; 1 \xrightarrow{\beta} \ldots 3X$$

$$X = D\text{-xylose}$$

Xylan in Rhodophyceae (Rhodymenia palmata), comprises 80% of β-1,4'-linked units. There is a structural similarity between the β-1,4'-linked xylans of the land plants and the xylans of the species of the green algae.

Rhodymenia xylan:

$$X1 \underline{\quad\beta\quad} 4 \ X \ 1 \underline{\quad\beta\quad} 4 \ X \ 1 \underline{\quad\epsilon\quad} \ \ldots \ 3 \ X \ 1 \underline{\quad\beta\quad} \ \ldots$$

## Mannans

Algal mannans are similar to those in ivory nut and iris although algal mannans probably do not contain galactose units (Percival, 1964). Mannans have been isolated from Porphyra umbicalis and species of Codium (C. fragile, C. latum, C. intricata, C. adhaereus), Acetabularia calyculus and Halicoryne wrightii. They are β-1,4'-linked and often represent up to ~60% of the crude fiber of the cell wall (Miwa et al., 1961).

## Algal Starches

Starches are the main storage products of some algae. True starches consisting of two principal constituents, i.e. amylose and amylopectin, have been found mainly in Chlorophyceae. The algal amyloses and amylopectins resemble those of plants (J. Love et al., 1963). Floridean starch consisting entirely of glucose residues is structurally different from starch from higher plants. Floridean starch is a typical photosynthetic product of certain red algae (Rhodophycophyta), however it is more closely related to the plant amylopectins than to the animal glycogens. Likewise it does gelatinize in water upon prolonged boiling (Meeuse et al., 1960).

Myxophycean starch generally occurs in the Myxophyceae. A similar polysaccharide has been isolated from blue-green algae (Cyanophyte starch) (Fredrick, 1959). Paramylum, the characteristic reserve product of the Euglenineae, yields mostly glucose upon hydrolysis and is closely related to starch.

Starch and related polymers appear to be absent in the family of Phaeophyceae. Table 1 shows the main storage products of algae (Fogg, 1953).

## Minerals and Elements

The word kelp itself originally referred to the burnt ash of the brown seaweeds. Subsequently the term became applied to the actual weeds. Ash from seaweed has been used for a long time as a source of soda. Algae also contain iodine and potash and depending on the ashing procedure, ammonia, tar or charcoal can be recovered.

Table I.   Biochemical Characteristics of the Algal Classes

| Class | Reserve products | Cell wall constituents | Sterols |
|-------|------------------|------------------------|---------|
| Chlorophyceae | starch<br>fat | cellulose<br>pectin | sitosterol<br>fucosterol<br>chondrillasterol<br>ergosterol |
| Chrysophyceae | leucosin<br>fat | pectin<br>silica | fucosterol<br>unidentified<br>  sterols |
| Xanthophyceae | fat | pectin<br>silica<br>cellulose<br>chitin(?) | sitosterol |
| Bacillario-<br>  phyceae | fat<br>leucosin(?) | silica<br>pectin | fucosterol<br>unidentified<br>  sterols |
| Cryptophyceae | starch | cellulose(?) | – |
| Dinophyceae | starch<br>fat | cellulose<br>pectin | – |
| Euglenineae | paramylum | none | – |
| Phaeophyceae | mannitol<br>laminarin | algin<br>fucoidin<br>cellulose | fucosterol |
| Rhodophyceae | floridoside<br>mannoglycerate<br>floridean<br>  starch | polygalactose-<br>sulphate esters<br>cellulose | fucosterol<br>sitosterol<br>unidentified<br>  sterols |
| Myxophyceae | myxophycean<br>  starch<br>cyanophycin | pectin<br>cellulose | none |
| Higher plants | starch<br>fat | cellulose<br>pectin<br>lignin | sitosterol<br>chondrillasterol<br>stigmasterol |

(After Fogg, 1953)


## Soda and Potash

The industry based on soda and potash recovery from algal ash has operated with alternating periods of prosperity and depression since the mid of eighteenth century.  The French were the first to use the ash of burnt seaweeds as a source of salt and soda, however, other European countries especially Ireland and Scotland picked up the technique quickly.  Algae primarily used in Europe were the species of Laminaria, Fucus and Ascophyllum nodosum.

Large areas on the Pacific coast of North America and Canada
are occupied by laminarians, e.g. Macrocystis, Nereocystis, and
Alaria which are used as a source of potassium carbonate.  The
average dry weight content in Nereocystis is around 19%, in
Macrocystis around 16% and in Alaria around 7%.  The amount varies
with season, habitat and wave activity.

When the weed is collected it is allowed to air dry and then
burned in cylindrical kilns.  Soda and potash are recovered from
the ash cake which accumulates.  Later when iodine was needed the
technology was modified to avoid the losses of iodine and this by-
product was produced.  Ash of these algae makes up 20% of the dry
weight.

Kelp is used mainly as a source of soda and potash in the
manufacture of soap, glassware, alum, etc.


Iodine

In 1811 Courtois discovered the new element iodine in the
ash of brown seaweeds and for 30 years kelp was the primary source
of this element.  Seaweed is about 1000 times richer than other
plants in iodine (Klincare, 1967).  Phaeophyta are particularly
rich in iodine.  In Laminariales and Desmarestiales the concentration may occasionally reach 30,000 times that in sea water, and
can constitute 1% of dry weight (Black, 1948).  The iodine is
typically present in iodide form (Shaw, 1962; Meguro, et al, 1967
Rhodophyta, Ptilota and some Chlorophyta, notably the culture
Codium intricatum, also contain an appreciable quantities of iodine.
The concentration of iodine within the algal cell is often followed
by using the radioactive isotope $I^{131}$ (Roche and Andre, 1962).
Iodine content varies with the season and the species of algae
studied.  The level of iodine in young plants is generally higher
than in old ones.

Iodine from kelp is still produced in Japan and it represents
about 5-7% of the world production.  Laminaria, Fucus, Ecklonia
and Eisenia are the genera principally employed.  In order to
improve the yield a new process has been designed to recover the
iodine from algae.  After drying the weed in rotary driers the dry
tissue is processed either by destructive distillation or by fermentation.  Temperature in the first is not very high in order to
prevent volatilization and losses of iodine.  Other by-products of
this process are ammonia, tar, as well as gas and charcoal.  The
fermentation process is far more elaborate and yields a greater
variety of products (Chapman, 1962) as is shown in Figure 6.

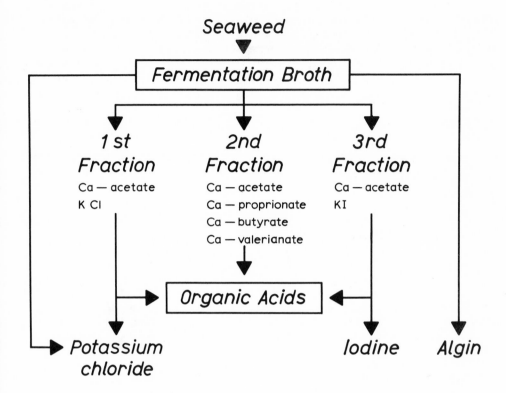

Fig. 6.  Iodine, potassium chloride and algin recovery from seaweed.


## Trace Elements

In addition to the halogens such as bromine and chlorine many minerals are concentrated in algae.  The use of seaweed as a source of trace elements is not new.  They have been used for their trace elements to supplement basic fodders and fertilizers.  Elements as copper, iron, zinc, cobalt, vanadium, molybdenum manganese, boron, aluminum and chromium, are found in seaweeds in sufficient trace quantities to make their use effective (Booth, 1964; Yamamoto et al., 1956).


## Human Food

In the Far East seaweeds have been used for centuries as a human food.  In Japan _Porphyra_ is widely cultivated for culinary

purposes.  It is recovered from bamboo poles inserted into shallow
waters.  When mature the alga is removed from the poles, dried and
pressed into sheets where it is known as Asakusa-nori.  Another
algal food called kombu is prepared from large laminarians, especially Laminaria, Alaria and Arthrothamnus.  The amount harvested
is considerable and runs over half million tons of wet weed per
year.  Other foods from known seaweeds are Wakame from Undaria
pinnatifida, Arame from Eisenia bicyclis, Hijiki from Hijikia
fusiforme, Miru from the green Codium and a variety of other weeds.

In the Pacific Islands raw algae of the Rhodophyta, Chlorophyta
and Phaeophyta are added to many common foods.  At least 70 species
are in common use.  Undaria and Sargassum as well as Caulerpa
racimosa are often used in the Philippines.

In South America dried and salted species of Ulva and Durvillea
are used in a food called "cachiyugo".  In Europe meals from seaweeds have never achieved wide popularity but in some countries
they are used to a limited extent.

Some attempts have been made to assess the nutritional value of
these edible marine algae.  Edible preparations of Nostoc from
China and Porphyra have high nitrogen levels.  Other food values
are (1) carbohydrates, (2) iodine, (3) minerals, (4) vitamins and
(5) other essential growth factors.

Extensive studies especially with unicellular algae, particularly Chlorella, have shown that these microorganisms are very high
in proteins and possess considerable nutritional value.  Due to
their unicellular nature these algae are ideally suited for cultivation by fermentation processes in either fermentation vessels or
"open-air" lagoons.  World food shortages have created a need for
developing new nonconventional food sources.  Algal culture represents one way to alleviate the world-wide food and protein shortage.
Currently the cost of algal protein can hardly compete with conventional sources of protein, but further research in this field should
reduce the cost for algal protein synthesis significantly.  Active
research programs are being conducted in this field.  Thacker and
Babcock (1957) reported yields of Chlorella pyrenoidosa of 13 g
d.w./liter in a batch 12 day fermentation.  Estimated cost of cells
produced on this basis would be 50 cents/pound.  Tamiya (1957)
reported a considerably lower figure of about 26 cents/pound which
could be reduced further using continuous-flow large scale production.

Production of algae in pilot scale units is feasible.  The
yield of Chlorella obtained in a closed system was 15-20 g d.w./m$^2$
day and this yield still can be improved by adjusting the optimal
parameters for photosynthetic growth.  It is necessary to recognize
that the yield is controlled by numerous factors.  Under natural

conditions in an aquatic medium the limiting influence of light, $CO_2$ availability, turbulence, growth supporting chemical components, etc., restrict maximum photosynthetic yields to a considerable extent. The following modification of the Baule-Mitscherlich limiting factor equation is suggested (Verduin, 1962) as a model of the limiting-factor relationship:

$$Y = Y_{opt} (1-e^{-x})(1-e^{-y})(1-e^{-z})$$

where:

| | |
|---|---|
| $Y$ | = photosynthetic yield per $m^2$ of production area |
| $Y_{opt}$ | = yield obtainable if all the factors are present at optimal levels |
| $x,y,z$ | = factors influencing the photosynthetic process |

It can be used to compare primary production.

Many authors have analyzed the protein content and examined amino acid composition in various species. Results vary over a wide range of values (Fowden, 1962; Rao and Tipnis, 1964; Bailey and Staehelin, 1969; Clement et al., 1968; Lewis, 1962; Lewis and Gonzalves, 1962).

Higher values of protein are found in the Chlorophyceae (Chlorella pyrenoidosa, C. ellipsoidea, C. vulgaris: 40-50%; Ulva 20-30%)(Rao and Tipnis, 1964) and Rhodophyceae, than in the Phaeophyceae. The blue-green alga Spirulina maxima (Clement et al., 1968) is very rich in proteins (63-68%) and vitamins. It possesses carbohydrates at levels between 18-20% and 2-3% lipids. Under continuous culture it yields ~15 g d.w./day/$m^2$.

The cultures of Chlorella are reported to contain from about 10-25% carbohydrates and 8-30% crude fat.

Table II presents amino acid composition of Spirulina maxima (Clement et al., 1968) in comparison with other algal proteins (Fowden, 1962), a standard protein of F.A.O. and Baker's yeast protein.

From a nutritional point of view the vitamin content of some algal cultures is high which makes them even more valuable as a food (Jensen, 1964; Pratt and Johnson, 1965).

Table III shows vitamin content of Chlorella grown under some different autotrophic conditions in comparison with other sources of microbial protein.

Table II. The amino-acid composition of the bulk protein of various algae compared with that of standard protein FAO.

[Amounts are given as g. amino-acid N/100g protein N]

| AMINO ACID | FAO | Chlorella vulgaris | Ulva sp. | Tribonema alginate | Navicula pelliculosa | Laminaria sp. | Chondrus sp. | Microcystis (Diplocystis) aeruginosa | Anabaena cylindrica | Phormidium sp. | Spirulina maxima |
|---|---|---|---|---|---|---|---|---|---|---|---|
| Reference | 1 | 2 | 2 | 2 | | | | | 2 | 2 | 1 |
| Essential amino-acids | | | | | | | | | | | |
| Arginine | – | 15.8 | 7.5 | 15.9 | 9.2 | 16.1 | 10.2 | – | 11.7 | 9.2 | – |
| Histidine | – | 3.3 | 1.2 | 3.7 | 2.8 | 1.6 | 1.8 | – | 2.5 | 3.8 | – |
| Isoleucine | 4.2 | 3.5 | – | 4.1 | 3.5 | – | – | 2.2 | 3.9 | – | 6.03 |
| Leucine | 4.8 | 6.1 | 5.2 | 6.4 | 7.2 | 2.5 | 5.3 | 4.2 | 6.2 | 2.1 | 8.02 |
| Lysine | 4.2 | 10.2 | 0.0 | 9.0 | 8.3 | 0.0 | 4.0 | – | 6.6 | 0.0 | 4.59 |
| Methionine | 2.2 | 1.4 | 0.0 | 1.4 | 1.2 | 0.0 | 0.0 | 1.7 | 1.2 | 2.0 | 1.37 |
| Phenylalanine | 2.8 | 2.8 | 2.3 | 2.8 | 3.4 | 1.0 | 1.5 | 4.4 | 2.9 | 1.1 | 4.97 |
| Threonine | 2.8 | 2.9 | – | 4.0 | 4.2 | – | – | 3.2 | 5.7 | 1.6 | 4.56 |
| Tryptophane | 1.4 | 2.1 | 0.3 | 1.8 | 1.1 | 1.1 | 1.6 | – | 1.0 | 0.2 | 1.40 |
| Valine | 4.2 | 5.5 | 5.2 | 7.5 | 7.5 | 5.1 | 2.8 | 4.1 | 7.0 | 6.7 | 6.49 |

Table II. Cont..

| AMINO ACID \ SOURCE OF PROTEIN | FAO [1] | Chlorella vulgaris [2] | Ulva sp. [2] | Tribonema alginate [2] | Navicula pelliculosa | Laminaria sp. | Chondrus sp. | Microcystis (Diplocystis) aeruginosa | Anabaena cylindrica [2] | Phormidium sp. [2] | Spirulina maxima [1] |
|---|---|---|---|---|---|---|---|---|---|---|---|
| **Non-essential amino-acids** | | | | | | | | | | | |
| Alanine | – | 7.7 | 6.5 | 8.4 | 6.5 | 6.4 | 3.7 | 5.4 | 6.0 | 5.2 | – |
| Amide N | – | 6.1 | – | 6.5 | 7.1 | – | – | – | 8.0 | – | – |
| Aspartic acid | 4.2 | 6.4 | 4.1 | 5.1 | 6.4 | 1.9 | 2.5 | 4.6 | 6.9 | 0.9 | 1.80 |
| Cystine | – | 0.2 | 1.8 | – | – | 3.4 | 1.6 | – | – | 0.0 | – |
| Glycine | – | 6.2 | 0.8 | 6.2 | 6.1 | 2.7 | 2.1 | 4.9 | 5.5 | 1.6 | – |
| Glutamic acid | – | 7.8 | 7.6 | 4.6 | 4.9 | 7.3 | 8.2 | 6.5 | 5.6 | 4.4 | – |
| Proline | – | 7.2 | 7.0 | 6.1 | 6.2 | 7.6 | 7.1 | 3.2 | 5.0 | 7.0 | – |
| Serine | – | 3.3 | – | 2.4 | 4.2 | – | – | 3.3 | 2.4 | – | – |
| Tyrosine | 2.8 | 2.8 | 0.0 | 3.0 | 1.9 | 1.9 | 2.3 | – | 1.6 | 1.8 | 3.95 |
| Total N | 33.5 | 101.3 | 49.5 | 98.9 | 9.17 | 58.6 | 54.7 | 47.7 | 89.7 | 46.0 | 53.18 |

References:

1) Clement et al., 1968.

2) Fowden, 1962.

Table III.   Vitamins in Algae [mg/kg]

|  | Chlorella pyrenoidosa (25°C) | Chlorella pyrenoidosa (outdoor pond) | Yeasts |
|---|---|---|---|
| Reference | Combs, 1952 | Yakult, 1960 | Handbook of biolog. data, ed. Saunders, 1960 |
| Ascorbic acid | – | 1800–4000 | 0 |
| Thiamine ($B_1$) | 9.5 | 3.7– 22.0 | 27–82 |
| Riboflavin ($B_2$) | 33.0 | 19.5– 55.0 | 33–80 |
| Niacin | 220.0 | 110.0–218.0 | 173–2670 |
| Panthotenic acid | 18.5 | – | 107–180 |
| Cholin | 2800.0 | – | 2100–3350 |
| Pyridoxin ($B_6$) | 21.5 | 10.0– 26.0 | 15–58 |
| Cobalamin ($B_{12}$) | 0.021 | – | – |

Marine algal culture provides a distinct advantage over culture of terrestrial plants, because desirable algal forms have a high protein content and can be cropped continuously. In large scale algal culture a fraction of the crop can be harvested daily and the algae production can be maintained at peak levels. On the basis of protein content the algae have a 3-fold advantage over leguminous plants. Their caloric yield per unit of environmental area under natural large scale culture is already equal to that of our best terrestrial crop plants at their season of peak production.

Continuous cropping of algae will, therefore, yield a 3-9 fold higher caloric conversion per year if cultures are maintained at peak efficiency. A 10-50-fold advantage in protein yield can be expected.

If man is capable of adapting to direct consumption of algal carbohydrate and protein (Rogers, 1963), a large proportion of earth's watery surface could be used for algae cultivation. It is estimated (Verduin and Schmid, 1966) that the efficienty of such a conversion would support a world population of the order $10^{13}$ people, or 3000 times the present world population. However, environmental problems would be created so that this figure is quite hypothetical. It does show the potential of algae in resolving the world nutrition problem.

Pigments

There are numerous algal pigments. The subject is too complex and voluminous to cover herein, however, some mention must be made of the carotenoids.

The carotenoids of algae include a variety of both common and unusual types (Egger, 1967; Goodwin, 1966). Their main importance is assessed in their value as provitamin A.

Carotenoids are yellow, orange, or red pigments of aliphatic or alicyclic structure composed of isoprene units (Gerosa, 1966) linked so that the methyl groups nearest the center of the molecule are in 1,5-position while all other lateral methyl groups are in 1,6-position. In general, natural carotenoids may be divided into two classes: 1) oxygen-free hydrocarbons, the carotenes
                        2) their oxygenated derivatives, the xanthophylls.

The principal carotene is β-carotene, although there are exceptions where α-carotene predominates (Nakayama, 1962). Table IV shows distribution of carotenoids and other pigments in the various classes of algae (Bogorad, 1962).

The recovery of carotenoids generally involves an extraction of the source material with acetone or methanol and/or some other organic solvent. Further separation procedures employ partition chromatography on silica gel or reverse-phase column chromatography using polyethylene and methanol. The pigments are then crystallized.

The other algal pigments are chlorophylls, tannins and phycobilins. Some species of Sargassum contain tannin which has been used in tanning industry (Ogino, 1962).


Fodder

For many years seaweeds have been used in maritime areas of the world as a stock food. Seaweeds have definite nutritional values, vitamins and other micronutrients. The major uses of brown and red algae have been for grazing of sheep, and to a lesser extent horses and cattle. There is a race of sheep on North Ronaldsay and in the Orkneys which lives entirely on seaweed for approximately 10 months per year. In Europe several factories have been built to manufacture stock feed using the brown rockweeds of Fucus, Ascophyllum and also some species of Laminaria. At present different kinds of supplementary seaweed fodder are on market. In these areas algae meal is used in place of hay, oats and potato tops. Laminarin appears to be the most digestable and therefore the time of harvesting is of major importance for obtaining maximum laminarin content.

Trace elements present in seaweeds are of a great importance to the sheep and cattle feeding on algae (Booth, 1964). Sheep have been shown to be very sensitive to an unbalanced copper content

Table IV.  Distribution of Chlorophylls, Biliproteins, and Carotenoids Among Algae

| Pigments | Charophyta | Chlorophyta — "Siphonales" [b] | Chlorophyta — Others | Euglenophyta | Xanthophyta | Chrysophyta | Bacillariophyta | Phaeophyta | Pyrrophyta | Cryptophyta | Cyanophyta | Rhodophyta |
|---|---|---|---|---|---|---|---|---|---|---|---|---|
|  | (1) | (2) | (3) | (4) | (5) | (6) | (7) | (8) | (9) | (10) | (11) | (12) |
| **Chlorophylls** | | | | | | | | | | | | |
| $a$ | + | + | + | + | + | + | + | + | + | + | + | + |
| $b$ | + | + | + | + | − | − | − | − | − | − | − | − |
| $c$ | − | − | − | − | − | ? | + | + | + | + | − | − |
| $d$ | − | − | − | − | − | ? | − | − | − | − | − | ± |
| $e$ | − | − | − | − | [a] | ? | − | − | − | − | − | − |
| **Biliproteins** | | | | | | | | | | | | |
| Phycocyanin | | | | − | − | − | − | − | − | +[b] | + | + |
| Phycoerythrin | | | | − | − | − | − | − | − | | + | + |
| **Carotenes** | | | | | | | | | | | | |
| α-Carotene | + | + | ±[c] | − | − | − | ± | ± | − | − | − | ± |
| β-Carotene | + | + | + | + | + | + | + | + | + | + | + | + |
| γ-Carotene | + | − | − | − | − | − | − | − | − | − | − | − |
| Lycopene | − | − | − | − | − | − | − | − | − | − | − | − |
| ε-Carotene | − | − | − | − | − | − | − | − | − | − | − | − |
| Unknown | − | − | − | + | − | − | + | − | − | + | − | − |

Table IV. Cont.

| | (1) | (2) | (3) | (4) | (5) | (6) | (7) | (8) | (9) | (10) | (11) | (12) |
|---|---|---|---|---|---|---|---|---|---|---|---|---|
| Xanthophylls | | L, N,S Sx V,Z | As L N V Z | As L, N, Un | | F L Un | Dd Dt F | Flx, F, L, V | Dd Dn P | ?Z | Apn, Apl Fle, ±l Mn, Ml, O, Z | T, ±Z 1, |

(After Bogorad, 1962)

Key

| | | |
|---|---|---|
| + | = | Present |
| − | = | Absent |
| ? | = | Insufficient information |
| Apn | = | Aphanicin |
| Apl | = | Aphanizophyll |
| As | = | Astaxanthin (euglenarhodone) |
| Dd | = | Diadinoxanthin |
| Dt | = | Diatoxanthin |
| Dn | = | Dinoxanthin |
| Fle | = | Flavacin |
| Flx | = | Flavoxanthin |
| F | = | Fucoxanthin |

| | | |
|---|---|---|
| L | = | Lutein |
| Mn | = | Myxoxanthin (aphanin, echinenone) |
| Ml | = | Myxoxanthophyll |
| N | = | Neoxanthin |
| O | = | Oscilloxanthin |
| P | = | Peridinin (sulcatoxanthin) |
| S | = | Siphonein |
| Sx | = | Siphonoxanthin |
| T | = | Taraxanthin |
| V | = | Violaxanthin |
| Z | = | Zeaxanthin |
| Un | = | Unknown |

a  Only observed in Tribonema bombycinum
b  Reported also in Cyanidium caldarium
c  Reported in Palmellococcus miniatus
d  Order Codialles and Derbesiales Feldmann

in the fodder and this can lead to an extremely high death rate
especially for the newly born and young lambs (Booth, 1964;
Hallsson, 1964). The effect of minute amount of copper in the
seaweed is more effective than a large dose of inorganic salt in
the form of copper sulphate. This results in an increase in the
butter-fat content in milk of lactating animals.

Seaweed has appeared to be the only good reliable source of
cobalt which is about ten times higher than in grass. Proper
levels of this element in the fodder is very important, overdosing
can lead to a serious disorder of metabolism. In the USA the
cobalt content in the fodder is restricted.

Fresh Laminariaceae were successfully preserved by ensiling
and this process is reported to give a wholesome and palatable
fodder for sheep. Fresh _Alaria_ was also dried after washing in
the rain and put in between layers of hay and thus given to cattle
and sheep with encouraging results. _Sargassum_ species are used
as fodder in China.

Eggs, from hens fed on seaweed meal, have an increased iodine
content.

Many fish, both marine and freshwater feed on planktonic or
attached type algae. Diatoms are apparently easily digested by
most fish, although the silica frustules are not utilized.

### Fertilizer and Manure

Seaweeds and algae collected along the shoreline have been
used for at least 100 years for fertilizer. Manurially the algae
used have a good nitrogen and potash contents but they are low
in phosphorus having only about one-third of the farm-yard manure
phosphate content. The nitrogen is not freely available and requires
time to pass into the soil. This makes seaweed manure a slow but
long acting fertilizer. Seaweed addition is well suited to light
sandy soils, which are generally deficient in potassium. The
physical condition of these light soils also improves (crumb structure) on account of the gelatinous nature of seaweed. This is
attributed to the high content of polysaccharides and its conse-
quent capacity for holding water (algin etc.).

Excesses of some salts in seaweeds may be compensated by com-
posting before use. However, the amount of common salt present is
not as excessive as earlier anticipated. Seaweed manure is particularly valuable because of the trace elements that it contains.
The total quantity of trace metals removed annually by a good crop
of grass could be replaced by the following quantities of seaweed
(Booth, 1964):

|          | per acre  |            | per acre |
|----------|-----------|------------|----------|
| iodine   | 4.5 lb.   | cobalt     | 4 cwt    |
| vanadium | 40 lb.    | zinc       | 20 cwt   |
| iron     | 2 cwt     | molybdenum | 26 cwt   |
| boron    | 2.5 cwt   | manganese  | 84 cwt   |
| copper   | 3 cwt     |            |          |

Soil itself can usually supply most of the elements and it is probable that under most conditions 1-2 cwt/acre/annum would be adequate. Soil pH often makes some of the elements unavailable to the plant. More data in this field as well as evaluation of toxic effects of some elements (boron, copper) are needed. For acid soils a useful dressing is provided by some Corallinaceae and Lithothamnion which contain a high proportion of lime (Walter-Levy, 1961).

Drying of the seaweed manure before application makes it more effective because when stored wet some of valuable materials such as potash are leached from it. Dry seaweed fertilizers of many kinds are available on the market. They are easily applied by mechanical distributors. The most economical way of applying seaweed is as a liquid extract (Milton, 1964). This is compatible with other liquid sprays and can be combined with various additives to achieve proper activity. Absolute safety is another advantage of these products. Liquifaction of the seaweed will bring about a product which gives the same result as composted seaweed. In this process the seaweeds are reduced to liquid form by pressurized alkaline hydrolysis. A wide selection of weeds is used for this purpose and those containing high percentage of laminarin, fucoidin and alginic acid are treated with mild alkali so that polysaccharides are not broken down to monomeric units. The final product which is black in color may be applied to the soil in a much diluted form of 1:500. The net effect of this substance is similar to the controlled application of small amounts of chemical fertilizer when applied over a prolonged time period since it enhances nitrogen availability and the release of bound phosphate and potach from the soil (Myklestad, 1964).

There is another aspect of the use of algae as fertilizer in connection with the nitrogen fixation properties of certain blue-green algae (Tabenkin, 1968). There are about 50 species of Cyanophyta which actively fix atmospheric nitrogen (Watanabe, 1960). In some rice crops inoculated with Tolyphotrix tenuis production increased 20% over the uninoculated controls. Seeding the rice fields with nitrogen fixing algae is of great economic importance. Large scale pure culture cultivations of the inoculum for rice fields is required. Controlled heterotrophic growth of these algal cultures would enable one to produce the symbiotic cultures in large volume cheaply (Watanabe and Yamamoto, 1967).

Antibiotics and Medicinal Uses

There is much evidence suggesting that certain species of
algae liberate extracellular products which inhibit growth of
other organisms as well as themselves.  The first observation
concerning antibiotic active substances secreted by algae was
given by Harder in 1917.  Since the works of Pratt et al. (1940)
on Chlorella vulgaris, many algal substances have been shown to
have bactericidal or bacteriostatic properties toward various
pathogenic species (Zaehmer, 1962; Walters, 1964).  Sensitivity
of diphteria, typhoid, paratyphoid and disenteric microorganisms
to a substance consisting of lipids, called cyanophycin, has been
reported (Mamaichuk and Matus, 1959), as well as marinamycin
(Soeda, 1962).

Pelvetia, Halidrys, Laminaria digitata and Polysiphonia has
been shown to possess antibacterial properties associated with
chlorophyll (Chesters and Stott, 1955).  Also among other species
antibacterial activity is often associated with photosynthetic
activity e.g. Nitzchia palatea.  An antimicrobial substance isolated from Liguharia tussilaginea obtained after steam distillation proved to be very efficient against Staphylococcus aureus
(dilution 1:5000) and Trichophyton asteroides (1:10000) (Kosuge
et al., 1963).  Some of the pronounced algal antibiotic agents
have been suggested to be oxidation products of unsaturated fatty
acids.

None of the extracellular antibiotic agents from algae has
proved to be of economic value in competing with fungal antibiotics in large-scale production.  They remain of academic interest
only.  However, since some green and other algae are common on
the sand of water works filters they probably decrease the bacterial count by their antibiotic action.  No doubt many other
antibiotics from algae will be isolated and identified.

The field of research in phycotherapy is young and the
results obtained are encouraging.  Some clinical assays have been
done for instance on wound healing.  Treatments applied to various
animals have been carried out and have given unexpectedly good
results.  In human therapeutics a certain tendency to return to
older methods of phycotherapy is arising (Lefèvre, 1964).

Some algae (e.g. Digenia simplex) have been used as an anti-
helminthics.  Sodium laminarin sulphate and fucoidin which are
similar to heparin suggest their possible use as blood coagulants.
This has been experimentally confirmed.

### Diatomite (Kieselguhr)

Diatomite is an inorganic material of algal origin containing about 86-88% silica. It gives extremely low losses of volatiles (4%) on ignition. It is almost chemically inert when processed. It is scooped-up from large deposits of diatom frustules of marine or freshwater origin which were layed down during Tertiary and Quaternary geological times. The siliceous cell walls are relatively insoluble and hence these sediments accumulated being relatively uncontaminated by clays, etc. Diatom walls from marine plankton contain 96.5% $SiO_2$ and only approximately 1.5% $Al_2O_3$ or $Fe_2O_3$. It has been found that the silica walls of diatoms adsorb positively charged colloids to the same degree as do dried silica gels. Diatomite has a high degree of porosity and surface area ranging around 120 $m^2$/g which is about five times the "visible" surface (Lewin, 1962). Processed diatomite earth is used mainly as a filtration aid in sugar refining process in the brewing industry, in wine making, and in antibiotic production for mycelium removal.

Another use of diatomite is as a filler for paints varnishes and paper products, as well as for insulating purposes where extreme temperatures are used. It is also used as a chemical catalysts.

Historically diatomite was used in A.D. 532 in Constantinople to make lightweight bricks used for building the Cathedral of St. Sophia. Alfred Nobel used diatomite to adsorb nitroglycerin manufacturing dynamite. Other materials are used for this purpose today. Annual sales in the USA show a steady increase exceeding 3000,000 tons per year (Round, 1965).

### References

Anderson, N.Y., T.C.S. Dolan, A. Penman, D.A. Rees, G.P. Mueller, D.J. Stacioff, N.F. Stanley, 1968. Carrageenans, IV. Variations in the structure and gel properties of κ-carrageenins, and the characterization of sulfate esters by IR spectroscopy. J. Chem. Soc., C 1968, 602-606.

Arai, K., 1961. Constituents of marine algae. VI. Agar of Ahnfeltia plicata. Nippon Kagaku Zasshi, 82, 771-774.

Araki, C., 1965. The polysaccharides of agarophytes. Proc. Int. Seaweed Symp., 5th, Halifax, N.S., 3-17.

Bailey, R.W., L.A. Staehelin, 1969. Chemical composition of isolated cell walls of Cyanidium caldarium. J. Gen. Microbiol., 54, (2), 269-276.

Black, W.A.P., 1948. The season variation of some of the sublittoral seaweed common to Scotland. J. Soc. Chem. Ind. (London), 67, 165-176.

Bogorad, L., 1962. Chlorophylls. Physiology and Biochemistry of Algae, R.A. Lewin, ed., Academic Press, New York, N.Y. 385–404.

Booth, E., 1964. Trace elements and seaweeds. Proceedings of the IV Int. Sympos. on Seaweeds, DeVirvile, A.D. and J. Feldman, ed., The Macmillan Co., London, England.

Chapman, V.J., 1962. The Algae. MacMillan and Co. Ltd., London, England.

Chesters, C.G.C., J.A. Scott, 1955. Production of antibiotic substances by seaweeds. Int. Seaweed Symp., 2nd, Trondheim, 49–53.

Clement, G., M. Rebeller, and P. Trambouze, 1968. Utilization massive du gaz carbonique dans la culture d'une nouvelle algae alimentaire. Revue de l'Institute Francais du Pétrole et Annales des Combustibles Liquides, Paris, France, 23, 702–711.

Clingman, A.L., J.R. Nunn, 1959. Red-seaweed polysaccharides. III. Polysaccharides from Hypnea specifera. J. Chem. Soc. 493–498.

Combes, G.F., 1952. Algae (Chlorella) as a source of nutrients for the chic. Science, 116, 453–454.

Cote, R.H., 1959. Disaccharides from fucoidin. J. Chem. Soc., 1959, 2248–2254.

Czapke, Karol, 1961. Agar-Agar from domestic sources. Przemysl Spozywczy, 15, 652–657.

Dillon, T., 1964. The laminarans. Proceedings of the IV Int. Sympos. on Seaweeds, DeVirvile, A.D. and J. Feldman, ed., The Macmillan Co., London, England, 301–305.

Doshi, Y.A., P.S. Rao, 1968. Biosynthesis of alginic - $C^{14}$ acid (G.). J. Label. Compounds, 4, 192–193.

Egger, K., 1967. Review of distribution and function of vegetable carotenoids. Biol. Rundsch., 5, 112–124.

Fogg, G.E., 1953. The Metabolism of Algae. John Wiley and Son, New York, New York.

Fogg, G.E., 1966. The extracellular product of algae. Oceanogr. Mar. Biol., 4, 195–212.

Fowden, L., 1962. Amino Acids and Proteins. Physiol. Biochem. Algae, A. Lewin, ed., Academic Press, New York, N.Y., 189–206.

Fredrick, I.F., 1959. Chromatographic patterns of polysaccharide-synthesizing enzymes. Physiol. Plantarum, 8, 936–944.

Frei, Eva and R.D. Preston, 1962. Configuration of alginic acid in marine brown alga. Nature, 196, 130–134.

Gerosa, V., 1966. Chemical comparison of algal carotenoids. Studi Trentini Sci. Natur., 43, 159–171.

Goldstein, N., F. Smith, A.M. Unrau, 1959. Laminarin. Chem. and Ind., 124, 881.

Goodwin, T.W., 1966. Carotenoids. Comp. Phytochem., 1966, 121–37.

Gryuner, V.S., 1961. Chemical and physicochemical characterization of agaroid, a constituent of the alga Phyllophora nervosa. Colloq. Intern. Centre Natl. Rech. Sc., No. 103, 183–190.

Gryuner, V.S., L.A. Evmenova, 1961. Black Sea algae Phyllophora as raw material for food and fodder. Rybn. Khoz., 37, (10) 61-62.

Gusev, M.V., 1961. Blue-green algae. Mikrobiologiya, 30, 1108-28.

Hallsson, S.V., 1964. The uses of seaweeds in Iceland. Proceedings of the IV Int. Sympos. on Seaweeds, DeVirvile, A.D., J. Feldman, ed., The Macmillan Co., London, England, 398-405.

Harder, R., 1917. Ernährungsphysiologische Untersuchungen an Cyanophyceen, hauptsächlich dam endophytischen Nostoc punctiforme. Z. Bot., 9, 145.

Haug, A., B. Larsen, 1964. Studies on composition and properties of alginates. Proceedings of the IV. Int. Symp. on Seaweeds, DeVirvile, A.D., J. Feldman, ed., The Macmillan Co., London, England.

Haug, A., 1965. Alginic acid extraction. Norw. 105,975 (Cl. C 07 c3), Feb. 6, 1965.

Holm-Hansen, O., 1968. Ecology, physiology and biochemistry of blue-green algae. Ann. Rev. Microbiol., 22, 47-70.

Jensen, A., 1964. Ascorbic acid in Ascophyllum nodosum, Fucus seratus, and Fucus vesiculosus. Proceedings of the IV. Int. Symp. on Seaweeds, DeVirvile, A.D., J. Feldman, ed., The Macmillan Co., London, England.

Kappanna, A.U., A.V. Rao, 1963. Preparation and properties of agar-agar from Indian seaweeds. Indian J. Technol., 1, (5), 222-4.

Kim, Ch.S., H.J. Humm, 1965. Red alga Gracilaria foliifera with special reference to the cell wall polysaccharides. Bull. Marine Sci., 15, (4), 1036-50.

Klincare, A., I. Strazda, I. Streipa, 1967. Iodine level in fruits and in some biologically active substances. Latv. Lauksaimn. Akad. Raksti, 20, 323-6, (Latvian).

Kosuge, T., M. Yokota, 1963. An antimicrobial substance isolated from Liguharia tussilaginea. Yakugaku Zasshi, 83, 422-3.

Kroes, H.W., 1968. Excretion of mucilage and other substances by brown algae of the tidal zone. Kon. Ned. Akad. Wetenschap. Versi. Gewone Vergad. Afd. Natuurk, 77 (10), 159-69.

Lefevre, M., 1964. Medicinal uses of algae. Algae and Man. D.F. Jackson, ed., Plenum Press, New York, N.Y.

Leon de, A.I., N. Eufemio, M. Pineda, 1963. Chemical composition of some Philippine algae. Philippine J. Sci., 92, 77-87.

Lewin, R.A., 1962. Proteins, peptides, and free amino acid contents in some species of Padina from southeastern coast of India. Current Sci., 31, 90-92.

Lewis, E.J., E.A. Gonzalves, 1962. Protein, peptide and free amino acid contents of some species of marine algae from Bombay. Ann. Botany, 26, 301-16.

Love, J., W. Mackie, J.W. McKinnell, E. Percival, 1963. Starch-type polysaccharides isolated from the green seaweeds Euteromorpha compressa, Ulva lactuca, Cladophora rupestris, Codium fragile and Chaetomorpha capillaris. J. Chem. Soc., 4177-82.

Mackie, I.M., E. Percival, 1959. Constitution of xylan from the
    green seaweed Caulerpa filiformis. J. Chem. Soc., 1151-6.
Mamaichuk, M.I., A.G. Matus, 1959. Sensitivity of diphteria,
    typhoid, paratyphoid and disenteric microorganisms to
    cyanophytin. Pyatigorsk. Farm. Inst., 3, 20-25.
Meeuse, B.J.D., M. Andries, J.A. Wood, 1960. Floridean starch,
    J. Exptl. Bot., 11, 129-140.
Meguro, H., T. Abe, T. Ogasawara, K. Tuzimura, 1967. Analytical
    studies of iodine in food substances. I. Chemical form of
    iodine in edible marine algae. Agr. Biol. Chem., 31, (9),
    999-1002.
Milton, R.F., 1964. Liquid seaweed as a fertilizer. Proceedings
    of the IV Int. Sympos. on Seaweeds, DeVirvile, A.D., J.
    Feldman, ed. The Macmillan Co., London, England, 428-431.
Ming - How Chi, 1962. Chemical composition of the Chinese econom-
    ic brown seaweeds, II. Hai Yang Yu Hu Chao, 5, (1), 1-10.
Miwa, J., Y. Iriki, T. Suzuki, 1961. Mannan and xylan as essen-
    tial cell wall constituents of some siphonous green algae.
    Colloq. Intern. Centre Natl. Rech. Sci., No. 103, 135-144.
Miyake, S., 1959. Distribution of alginate in the body of algae
    from the viewpoint of its extraction. Kogyo Kagaku Zasshi,
    62, 422-25.
Myklestad, S., 1964. Experiments with seaweed as supplemental
    fertilizer. Proceedings of the IV Int. Sympos. on Seaweeds.
    DeVirvile, A.D., J.Feldman, ed., The Macmillan Co., London
    England, 432-438.
Nakayama, T.O.M., 1962. Carotenoids. Physology and Biochemistry
    of Algae. R.A. Lewin, ed., Academic Press, New York, N.Y.,
    409-416.
Nishide, H., 1961. Chemistry and Industry of alginic acid. Nosan
    Kako Gijutsu Kenkyu Kaishi, 8, 149-57.
Nunn, J.R., J. Parolis, 1968. A polysaccharide from Aeodes
    orbitosa. Carbohyd. Res., 6 (1), 1-11.
Ogino, C., 1962. Tannins and vacuolar pigments. Physiology and
    Biochemistry of Algae, R.A. Lewin, ed., Academic Press,
    New York, N.Y., 437-443.
O'Neill, A.N., 1955. Derivatives of 4-O-β-D-galactopyranosyl-3.6.-
    anhydro-D-galactose from κ-carrageenin. J. Amer. Chem. Soc.,
    77, 6324-6.
Painter, T.J., 1960. The polysaccharide of Furcellaria fastigiata.
    I. Isolation and partial mercaptolysis of a gel fraction.
    Can. J. Chem., 38, 112-118.
Peat, S., J.R. Turvey, 1965. Polysaccharides of marine algae.
    Fortschr. Chem. Org. Naturstoffe, 23, 1-45.
Percival, E., 1964. Algal polysaccharides and their biological
    relationships. Proceedings of the IV Int. Sympos. on Sea-
    weeds, DeVirvile, A.D., J. Feldman, ed., The Macmillan Co.,
    London, England.
Percival, E., R.H. McDowel, 1967. Chemistry and Enzymology of
    Marine Algal Polysaccharides, Academic Press, New York, N.Y.

Pratt, R., J. Fong, 1940. *Chlorella* vulgaris.II. Evidence that *Chlorella* cells form a growth inhibiting substance. Am. J. Botany, 27, 431-436.

Pratt, R., E. Johnson, 1965. Production of thiamine, riboflavine, folic acid, and biotin by *Chlorella vulgaris* and *Chlorella pyrenoidosa*. J. Pharm. Sci., 54, (6), 871-4.

Rae, A.V., I.C. Mody, 1968. Extraction of alginic acid and alginates from brown seaweeds. Indian J. Technol., 3 (8), 261-2.

Ralph, B.J., V.J. Bender, 1965. Isolation of two new polysaccharides from the cell wall of *Polyporus tumulosus*. Chem. Ind. (London), 26, 1181.

Rao, V.S., U.K. Tipnis, 1964. Protein content of marine algae from Gujarat coast. Current Sci. (India), 33 (1), 16-17.

Roche, J., Y. Andre, 1962. Autoradiographic study of the fixation of radioiodine by marine algae. Compt. Rend. Soc. Biol., 156, 1968-71.

Rogers, A.R., 1963. Protein digestion and utilization *in vivo* studied with uniformly $C^{14}$- labeled algal protein. Univ. Microfilms (Ann Arbor, Mich.), Order No. 63-2907, 178 pp., Dissertation Abstr., 23, 3111.

Round, F.E., 1965. *The Biology of the Algae*. Edward Arnold Ltd. London, England.

Saunders, W.B., ed., 1956. *Handbook on Biological Data*. Saunders, W.B., Co., Philadelphia, Penn.

Schmid, O.J., 1962. Constituents of marine algae. Botan. Marina, Suppl. 3, 67-74.

Schmitt, M., 1961. The carrageenins. Colloq. Intern. Centre Natl. Rech. Sci., No. 103, 199-204.

Shaw, T.I., 1962. Halogens. *Physiology and Biochemistry of Algae*, R.A. Lewin, Ed., Academic Press Inc., New York, N.Y., 247-253.

Smith, R., R. Montgomery, 1959. *Chemistry of plant gums and mucilages*. Reinhold Publ. Corp., New York, N.Y.

Soeda, M., M. Mitomi, 1962. Marinamycin - Chemical Studies. J. Antibiotics (Japan), Ser. A, 15, 182-6.

Stanley, N.F., 1963. Treating polysaccharides of seaweeds of the Grigartinaceae and Solienaceae families. U.S. 3,094,517, June 18, 1963.

Steinmetz, C.P., V. Gunther, R. Hinterwaldner, 1969. Treating marine algae. Brit. 1, 154, 137 (Cl. C 086), 4 January, 1969.

Tabenkin, B., 1968. Algal fermentations. Paper presented at the Ann. Mtg. of the Am. Soc. for Microbiol., (Detroit, Michigan), May 5-10, 1968.

Tamiya, H., 1957. The mass culture of algae. Ann. Rev. Plant Physiol., 8, 309.334.

Tsapko, A., 1966. New raw materials for the production of sodium alginate. Rybn. Khoz., 42 (7), 65-6, (Russ.).

Usov, A.I., N.K. Kochetkov, 1968. Polysaccharides of red algae. II. Polysaccharides of red alga *Odonthalia corymbifera*. Separation of 6-0-methyl-D-galactose. Zh. Obshch. Khim., 38 (2), 234-8, (Russ.).

Verduin, J., 1962. Principles of primary productivity: Photo-synthesis under completely natural conditions. Algae and Man, D.F. Jackson, ed., Plenum Press, New York (1964).

Verduin, J., W.E. Schmid, 1966. Evaluation of algal culture as a source of food supply. Dev. In. Microbiol., 7, 205.

Walter-Levy, L., R. Strauss, 1961. The mineral constituents of calcareous algae. Colloq. Intern. Centre Natl. Rech. Sci., No. 103, 39-50.

Walters, B., 1964. Antibiotic and toxic substances from algae and mosses. Planta Med., 12 (1), 85-99.

Watanabe, A., 1960. Collection and cultivation of nitrogen fix-ing blue-green algae and their effect on the growth and crop yield of rice plants. Proc. Symp. Algology, Indian Council of Agric. Res., New Delhi., 162-166.

Watanabe, A., Y. Yamamoto, 1967. Heterotrophic nitrogen fixation by the blue-green alga, Anabaenopsis circularis. Nature, 214, 738.

Yakult, K.K., 1960. Chlorella, Honska Bolletin, Japan Chlorella Research Institute, Tokyo.

Yamamoto, T., T. Fujita, M. Ishibashi, 1956. Chemical studies on the ocean. XCV-XCVI. Chemical studies on seaweeds. Nippon Kagaku Zasshi, 86, (1), 49-59.

Yatsenko, G.K., 1962. Biochemical composition of Cystoseira bar-bata. Nauk. Zap. Odesk. Biol. St., Akad. Nauk. Ukr. RSR, 72-9.

Young, E.G., 1961. Carrageenin and related polysaccharide sul-fates. Colloq. Intern. Centre Natl. Rech. Sci., No. 103, 173-81.

Zaehner, H., 1962. Antibiotics. Fortschr. Botan., 24, 471-81.

# ALGAL TOXINS

Edward J. Schantz

Department of the Army

Fort Detrick, Frederick, Maryland

Several species of algae produce very potent poisons (Gorham, 1960; Shilo, 1967; Schantz, 1968).  When we propose to culture algae for food for man and animals we must also contend with those algae that produce toxins and may contaminate food cultures.  In certain areas of this country, in fact throughout the world, sickness and death have occurred in domestic animals that have consumed toxic algae by drinking polluted water from shallow fresh water lakes and pools (Hughes et al., 1958).  The areas affected in North America are mainly the shallow fresh water lakes of northern United States and southern Canada.  Similar outbreaks of toxic algae have been reported from Russia, South America, Australia and Africa. The organisms that have been identified as causing this type of poisoning in the lakes and pools are species of blue-green algae.

Until recently man has not considered seriously the direct consumption of algae as a source of food and therefore has not been confronted with the problem in this way.  However, man has been poisoned by algal toxins through the food chain by eating shellfish at certain times.  The classical example in this case is the relation of the toxic marine dinoflagellate, Gonyaulax catenella, to paralytic shellfish poisoning reported by Sommer and Meyer (1937) and Sommer et al. (1937) at the University of California.  Occasionally in local areas shellfish become extremely poisonous and have caused sickness and death in humans that have eaten the shellfish. Shellfish acquire this poison from certain dinoflagellates that grow in the water where the shellfish feed.  In general, the shellfish remain poisonous only when feeding on a sufficient number of the poisonous organisms.  Along the California coast mussels became too toxic for human consumption when 200 or more G. catenella cells were found per ml. of water.  Low counts such as this can only be

83

detected by microscopic examination of the water. This organism, like many others, will produce a visible "red tide" when the count reaches 20,000 or more per ml. These organisms usually bloom for two to three weeks and within a week or two after the bloom the shellfish are safe for human consumption again. The poison is bound in the dark gland or hepatopancreas of the shellfish and causes no observable disturbance to its physiology or appearance. Because of this fact there is no means of distinguishing poisonous shellfish from those that are safe to eat, except by a bioassay with mice or other suitable animals.

The regions of the world where paralytic shellfish poisoning sometimes called mussel poisoning, has occurred most often are around the North Sea, the north Atlantic coast of America, the north Pacific coast of America from central California to the Aleutian Islands, and the coastal areas of Japan and of South Africa. In general, these areas are 30° or more north or south latitude. Table 1 lists some of the toxic algae that have caused trouble and which will be discussed in this report. Much of the background information on algal toxins was presented two years ago at an American Chemical Society symposium on Microbial Toxins (Schantz, 1968). The purpose of this presentation is to review and bring the information up to date.

The first three species listed in Table 1 are mainly responsible for cases of paralytic shellfish poisoning. G. catenella occurs along the Pacific coast and has definitely been established as the cause of paralytic poison in California mussels and probably in Alaska butter clams. Gonyaulax tamarensis occurs along the North Atlantic coast of America and throughout the North Sea area where it has caused clams, scallops and mussels to become poisonous (Needler, 1949; Prakash, 1963). The most recent outbreak of shellfish poisoning caused by this organism occurred during the latter part of June and July of 1968 along the northeast coast of England (Wood, 1968; Robinson, 1968). Another closely related species, Gonyaulax acatenella, occurring along the coast of British Columbia has caused clams to become poisonous (Prakash and Taylor, 1966). This organism produces a poison similar to the above two organisms. Gonyaulax polyedra often blooms along the coast of southern California. Schradie and Bliss (1962) have reported that this organism produces a poison in the natural state that has some properties similar to that produced by G. catenella. However, when the organism does produce a poison, it probably does so only under a specific set of conditions and has never been known to cause shellfish to become poisonous. Laboratory cultures have not proven to be toxic. Because the organism grows abundantly, it has been suggested by Patton et al. (1967) that the cells may serve as an animal food. The essential amino acid composition resembles that of casein. Gonyaulax monilata often blooms to red tide proportions in the Gulf of Mexico and produces a poison that is toxic

Table 1.  Toxins Produced by Various Algae
Pyrrophyta; Dinoflagellates

| Species | Common Location | Biological Characteristics | Chemical Characteristics |
|---|---|---|---|
| Gonyaulax catenella | Pacific coast of North America | Toxic to all higher animals. Predators bind toxin. Poison blocks sodium influx in nerve cells. | Among most potent poisons. Low mol wt (372). Water-soluble tetra-hydropurine. |
| Gonyaulax tamarensis | North American Atlantic coast | Appears similar to G. catenella poison. | Appears similar to G. catenella poison. |
| Gonyaulax polyedra[1] | California coast | Toxic to mice and fish. Produces poison only under some conditions. | Somewhat similar to G. catenella poison. |
| Gonyaulax monilata | Gulf of Mexico | Toxic to fish. | Not characterized. |
| Gymnodinium breve | Gulf of Mexico | Toxic to fish and mice. May be several poisons. | One poison is lipid soluble. Mol wt 1500. |
| Gymnodinium veneficum | Isolated from the English Channel | Toxic to fish and mice. | High molecular weight. Water soluble. |

Table 1 Continued.  Toxins Produced by Various Algae
Cyanophyta; Blue-Green Algae

| Species | Common Location | Biological Characteristics | Chemical Characteristics |
|---|---|---|---|
| Microcystis aeruginosa | Shallow fresh water lakes | Toxic to many animals. (FDF)[2] | Cyclic polypeptide of 10 amino acid units. Mol wt about 1200. Water soluble. |
| Anabaena flos-aquae | Shallow fresh water lakes | Toxic to many animals. Kills faster than M. aeruginosa. (VFDF)[3] | Water soluble. |
| Aphanizomenon flos-aquae | Fresh water | Same as G. catenella | Same as G. catenella |
| Schizothrix calcicola[4] | Pacific Ocean | Somewhat like ciguatera poisoning in man. | Not characterized. |
| Caulepra (several species)[5] | Western Pacific Ocean | Dizziness in man. | Not characterized. |

Table 1 Continued.  Toxins Produced by Various Algae

Chrysophyta; Yellow-Brown Algae; Phytoflagellates

| Species | Common Location | Biological Characteristics | Chemical Characteristics |
|---|---|---|---|
| Pyrmnesium parvum | Brackish water ponds and estuarine water. | Toxic to fish and all gill-breathing animals.  Potent hemolytic agent. | Water soluble. Non-dialyzable thermolabile. May be a saponin-like compound. |

[1]This species has been proposed as a food (see Patton et al. 1967).

[2]FDF  Fast death factor (see Gorham 1960).

[3]VFDF  Very fast death factor (see Gorham 1960).

[4]This species produces a poison but its relation to ciguatera poisoning from eating certain fish is speculative (see Banner 1967).

[5]Several species of Caulepra, when eaten by man, cause dizziness, but the symptoms are not well defined.

to fish but not to chickens (Connell and Cross, 1950; Gates and
Wilson, 1960; Ray and Aldrich, 1967). Oysters do not filter and
feed on this organism. Abbot and Ballantine (1957) described a
poison from Gymnodinium veneficum that is toxic to fish and mice
and will cause bivalues to close and stop filtering. Within the
past several years the toxin from Gymnodinium breve has been described in more detail. Starr (1958) described the toxicity to
fish and Ray and Aldrich (1965, 1967) described the toxicity to
chicks. McFarren et al. (1965) described the occurrence of a
ciguatera-like poison in oysters from the Gulf of Mexico and in
G. breve cultures. Poisonous substances have been isolated in pure
form and partially characterized from cultured G. breve cells
(Spikes and Ray, personal communication). These poisons are more
soluble in lipid solvents than in water and have a molecular weight
of about 1500. The present information indicate that there may be
at least two toxins, one of which is an acetylcholine esterase
inhibitor (Martin and Chatterjee, 1969).

One of the first toxins isolated and characterized from the
blue-green algae was that from Microcystis aeruginosa, which, as
stated above, has caused poisoning in many animals. Bishop et al.
(1959) in Canada have characterized this toxin as a cyclic poly-
peptide containing 10 amino acid units with a molecular weight of
about 1,200. It is a water soluble substance. When the minimum
lethal dose of 0.5 mg of this toxin is injected intraperitoneally
into mice death occurs in 30 to 60 minutes. Another important
toxin from the blue-green algae is that from Anabaena flos-aquae
which is also extremely toxic to many farm animals and kills very
rapidly. In mice death occurs within 15 minutes or less depending
upon the dose and the action is very much like that of the shellfish poisons. This toxin was described by Gorham and coworkers
(1960) as the very fast death factor and has not been isolated in
pure form. Just recently Jackim and Gentile (1968) from West
Kingston, Rhode Island reported the isolation of a toxin from
Aphanizomenon flos-aquae which they claim to be similar to the
toxin isolated from shellfish. It may be similar to the substance
which the Canadians called the very fast death factor from Anabaena
flos-aquae.

Another toxin of great interest is one produced by one of the
yellow-brown alga, Pyrmnesium parvum. This alga grows in the
brackish water ponds and estuary waters and produces a toxin that
is lethal to fish and all gill breathing animals. The toxin produced by the organism inhibits the transfer of oxygen across the
gill membranes and has been a great problem in the commercial carp
tanks in Israel. Several investigators in Israel have isolated
and studied the toxin to a great extent (Otterstrøm and Steemann-
Nielsen, 1939; Reich et al., 1965; Parnas, 1963; Shilo and Rosenberger, 1964). It is non-dialyzable, thermolabile and believed to
be a saponin-like compound that is a potent hemolytic agent. As

far as I know it has not been obtained in purified form.

Although all of the algal poisons mentioned cause considerable economic losses, the ones that have caused human deaths and sickness and are of great public health importance in certain areas are the poisons from certain toxic algae that have caused shellfish poisoning in humans. The symptoms of shellfish poisoning in humans begins by a numbness in the lips, tongue and fingertips within 30 minutes after the misfortune of eating poisonous shellfish. Progressive paralysis follows and death results from respiratory failure in 2 to 12 hours depending on the dose. If a person survives 24 hours the prognosis is good and no lasting effects are apparent. The poison from poisonous California mussels and Alaska butter clams was isolated in pure form at my laboratory after several years of cooperative work with the University of California and Northwestern University (Schantz et al., 1957; Mold et al., 1957; Schantz et al., 1961). Poisonous mussels were collected along the Pacific coast near San Francisco and butter clams from southeastern Alaska. Acidified water extracts of the finely ground dark gland of the mussels and the siphons from clams were made. The purification of the poison in the extract was accomplished by ion exchange chromatography on carboxylic acid resins, followed by chromatography on acid washed alumina in absolute ethanol. By this procedure a white hygroscopic product was obtained that had a potency of 5,500 mouse units (MU) per milligram of solids. A mouse unit (MU) is defined as the minimum amount of poison, injected intraperitoneally, that will kill a mouse weighing 20 grams in 15 minutes. Further chromatography employing a variety of techniques failed to improve the potency or change any of its chemical and physical properties. The purified poison is a dibasic salt with a $pK_a$ at 8.2 and 11.5 and is very soluble in water. Its molecular weight as a dihydrochloride salt is 372. It has no ultraviolet absorption, gives positive Benedict-Behre and Jaffe tests similar to that given by creatinine and is completely detoxified by mild catalytic reduction with the uptake of 1 mole of hydrogen per mole of poison at atmospheric pressure. The purified clam poison was studied by Rapoport et al. at the University of California, Berkeley and in 1964 they proposed that its structure was an unusually substituted tetrohydropurine and named the substance Saxitoxin. The complete structure of this poison has not been fully worked out.

After we had purified the poison from toxic mussels and clams, studies were initiated to culture G. catenella and isolate the poison from this organism. An axenic culture of G. catenella was obtained through the courtesy of Luigi Provasoli, Haskins Laboratories, New York and used for these studies. This organism was cultured in sterile sea water at 13°C. After 17 days the count reached about 30,000 and the cells were filtered from the media and lysed with dilute HCl at pH 2-3. This acid extract of the cells was processed through the carboxylic acid ion exchange resins and acid washed alumina, exactly as the extracts of poisonous mussels and clams. The final product from this process was equal in toxicity to mice and identical in all of its chemical and physical properties to those of mussels and clam poisons.

Table 2 shows a comparison of the properties of the poison from cultured G. catenella cells with poison from mussels and clams. The properties are identical within experimental error. In an attempt to determine if there may be some differences in structure, Rapoport carried out degradation studies on each of the three poisons under identical conditions and compared the properties of each of the degradation products. Each degradation product was identical in ultraviolet and infrared absorption, nuclear magnetic resonance and in $R_f$ values on paper chromatography. From these data it was concluded that the structures of all three poisons must be identical (Schantz et al., 1966).

These studies have shown that the poison produced by G. catenella in its natural state in the Pacific Ocean was absorbed by the mussels with no change in the chemical structure. Because the poison was produced in axenic culture it is concluded that the poison is a product of the dinoflagellate and not due to a symbiotic effect of the bacteria that are normally associated with these dinoflagellates in the natural state (Burke et al., 1960).

These poisons are among the most potent known to man. In terms of the purified poison one MU is equal to 0.18 micrograms. The intravenous dose for a rabbit is 3-4 micrograms per kilogram of body weight. Accidental cases of poisoning along the California coast and in Canada have indicated that the lethal dose for man is around 20,000 MU and may be as low as 3000 MU in some cases. In terms of weight of poison this would amount to about 1/2 to 4 mg by oral dose. In experimental animals the oral dose is about 10 times the intraperitoneal dose. There is no known effective antidote for the poisons. However, the Klamath Indians used the gum from the sugar pine tree (Thompson, 1916) to overcome the toxic effects, but studies in my laboratory have not shown any of the products from the sugar gum tree to be effective against the poison in experimental animals. Because the diaphram mussels are particularly sensitive to the poison and death usually results from respiratory failure, artificial respiration has been used and believed to be effective to some extent, particularly in marginal cases.

These poisons have become of special interest to physiologists because they block the propagation of impulses in nerves and skeletal mussels without depolarization. Evans (1964, 1965) at the Sherrington School of Physiology in London, and Kao and Nishiyama (1965) at the New York State University studied the mechanism of action of the purified poison and found that the block is due to some specific interference with the increase in sodium permeability,

Table 2.  Comparison of Properties of Poison
from Cultured Gonyaulax catenella Cells
with Poison from Mussels and Clams

| Property | Clam Poison | Mussel Poison | G. catenella Poison |
|---|---|---|---|
| Bioassay (MU/mg)[a] | 5200 | 5300 | 5100 |
| Specific optical rotation | +128° | +130° | +128° |
| $pK_a$ | 8.3; 11.5 | 8.3; 11.5 | 8.2; 11.5 |
| Diffusion coefficient | $4.9 \times 10^{-6}$ | $4.9 \times 10^{-6}$ | $4.8 \times 10^{-6}$ |
| Absorption in ultraviolet and visible[b] | None | None | None |
| N content (Kjeldahl) | 26.8 | 26.1 | 26.3 |
| Sakaguchi[c] | - | - | - |
| Benedict-Behre[c] | + | + | + |
| Jaffe[c] | + | + | + |
| Reduction with $H_2$ | Dihydro derivative, nontoxic | Dihydro derivative, nontoxic | Dihydro derivative, non toxic |

[a]All bioassay values are within experimental error of the value
5500 ± 500 MU/mg solids reported previously for clam and mussel
poisons.

[b]Infrared absorption of G. catenella poison was identical with
that of clam and mussel poisons.

[c]Tests carried out as described by Mold et al. (1957).

Courtesy Am. Chem. Soc., Biochemistry, 5, 1191 (1966).

normally associated with excitation. They also found that the rest-
ing membrane conductances attributed chiefly to potassium and
chloride permeabilities are unaffected. The toxin produced by G.
catenella is unique in its physiological action, in that conduction
is blocked in dorsal and ventral root fibers in the cat by a concentration of $5 \times 10^{-9}$ grams per ml (Evans, 1968). Only one other
known substance, tetrodotoxin from the puffer fish, is equivalent
to it in this action. Tetrodotoxin is entirely different in its
chemical structure.

Except for the shellfish poisons, the basic mechanisms of
action of the various algal toxins have not been studied extensively
and much more work needs to be done in this respect. Abbot and
Ballantine (1957) found the poison from G. veneficum produced
depolarization of the membranes of nerve and skeletal muscle and
Sasner (1965) found the same results with the poison from G. breve.
Ulitzur and Shilo (1966) have studied the mechanism of action of
the poison from P. parvum and have found it to be complicated by
many cofactors bringing about changes in the gill membrane permeability. The fast death factor from M. aeruginosa causes cellular
breakdown in the liver of animals but the basic cause for this
action is not known. The very fast death factor from Anabaena flos-
aquae and the recently isolated poison from Aphanizomenon flos-aquae
probably has an action similar to that of the poison from G. catenella (Sawyer et al., 1968).

Although progress has been made on the elucidation of the
nature and properties of the shellfish poisons and the poisons from
the dinoflagellates, many problems still remain that are important
economically and academically. One of these problems concerns the
poison occurring in the Alaska butter clams. These clams are one
of the most palatable species known but in some regions they are
too poisonous for human consumption. Although the cause of the
toxicity in the clams is open to debate, there is some evidence
that it may be G. catenella or a closely related species (Schantz
and Magnusson, 1964). The cause and a means of controlling the
poison in these clams would be of great economic value to Alaska.
A study of the means by which shellfish bind the poison in their
own system might lead to a means of removing the poison from shellfish without destroying the commercial value.

Other studies of considerable interest would be the mechanism
by which the various species of dinoflagellates synthesize the
poison. G. tamarensis produces a very potent poison that appears
to be like G. catenella poison in its biological action, but indi-
cations are that it is somewhat different in its chemical and
physical properties. This poison has been purified recently in our
laboratory but no extensive studies have been carried out on its
chemical and physical properties. Other toxic organisms that could
cause poisoning have been encountered in recent years and need

further study. Recently Japanese workers, Konosu et al., (1968),
have reported a toxin in crabs which is very similar, if not identical, in chemical structure to the shellfish poisons and to the
dinoflagellate poison from G. catenella. This might indicate that
some organism upon which the crabs are feeding may be poisonous.
Ciguatara poisoning in humans from eating certain fish caught in
the south Pacific area particularly, is believed to be the result
of these fish feeding on marine algae (Helfrich and Banner, 1966;
Banner, 1967). Recently, certain species of Caulerpa, a marine
blue-green alga, have been found to contain toxic substances
(Doty, M.S. and G. Aguilar-Santos, personal communication).

The trend toward the use of more products from the sea as food
for man and the culturing of various algae for food makes it important that we know more about various toxic algae and the toxins
that they produce.

## REFERENCES

Abbot, B.C., and D. Ballantine. 1957. The toxin from Gymnodinium
    veneficum Ballantine. J. Marine Biol. Assoc. U.K. 36, 169-
    189.
Banner, A.H. 1966. Marine toxins from the Pacific. 1. Advances
    in the investigation of fish toxins. Animal Toxins 157-165.
    Editors, F.E. Russel and P.R. Saunders, Pergamon Press,
    Osford and New York 1967.
Bishop, C.T., E.F.L.J. Anet, and P.R. Gorham. 1959. Isolation and
    identification of the fast death factor in Microcystis
    aeruginosa NRC 1. Can. J. Biochem. Physiol. 37, 453-471.
Burke, J.M., J. Marchisotto, J.J.A. McLaughlin, and L. Provasoli.
    1960. Analysis of the toxin produced by Gonyaulax catenella
    in axenic culture. Ann. N.Y. Acad. Sci. 90, 837-842.
Connell, C.H., and J.G. Cross. 1950. Mass mortality of fish
    associated with the protozoan Gonyaulax in the Gulf of Mexico.
    Science 112, 359-363.
Evans, M.H. 1964. Paralytic effects of paralytic shellfish poison
    on frog nerve and muscle. Brit. J. Pharm. and Chemotherapy
    22, 478-485.
Evans, M.H. 1965. Cause of death in experimental paralytic shell-
    fish poisoning (PSP). Brit. J. Exptl. Pathol. 46, 245-253.
Evans, M.H. 1968. Topical application of Saxitoxin and tetro-
    dotoxin to peripheral nerves and spinal roots in cat. Toxicon
    5, 289-294.
Gates, J.A., and W.B. Wilson. 1960. The toxicity of Gonyaulax
    monilata to Mugil cephalus. Limnol. Oceanogr. 5, 171.
Gorham, P.R. 1960. Toxic waterblooms of blue-green algae. Can.
    Vet. J. 1, 235-245.

Helfrich, P., and A.H. Banner. 1968. Ciguatera fish poisoning 11. General patterns of development in the Pacific. Bishop Museum Occasional Papers 23, 371-382.

Hughes, E.O., P.R. Gorham, and A. Zehnder. 1958. Toxicity of a unialgal culture of Microcystis aeruginosa. Can. J. Microbiol. 4, 225-236.

Jackim, E., and J. Gentile. 1968. Toxins of a blue-green alga: Similarity to Saxitoxin. Science 162, 915.

Kao, C.Y., and A. Nishiyama. 1965. Action of Saxitoxin on peripheral neuromuscular systems. J. Physiol. (London) 180, 50-66.

Konosu, S., A. Inone, T. Noguchi, and Y. Hashimoto. 1968. Comparison of crab toxin with Saxitoxin and tetrodotoxin. Toxicon 6, 113-117.

Martin, D.F., and A.B. Chatterjee. 1969. Isolation and characterization of a toxin from the Florida red tide organism. Nature 221, 59.

McFarren, E.F., H. Tanabe, F.J. Silva, W.B. Wilson, J.E. Campbell, and K.H. Lewis. 1965. The occurrence of a ciguatera-like poison in oysters, clams and Gymnodinium breve cultures. Toxicon 3, 111-123.

Mold, J.D., J.P. Bowden, D.W. Stanger, J.E. Maurer, J.M. Lynch, R.S. Wyler, E.J. Schantz, and B. Riegel. 1957. Paralytic shellfish poison Vll. Evidence for the purity of the poison isolated from toxic clams and mussels. J. Am. Chem. Soc. 79, 5235-5238.

Needler, A.B. 1949. Paralytic shellfish poisoning and Gonyaulax tamerensis. J. Fisheries Res. Board Can. 7, 490-504.

Otterstrøm, C.V., and E. Steemann-Nielsen. 1939. Two cases of extensive mortality in fishes caused by the flagellate Pyrmnesium parvum Carter. Rept. Danish Biol. Sta. 44, 5-24.

Parnas, I. 1963. The toxicity of Prymnesium parvum (a review). Israel J. Zool. 12, 15-23.

Patton, S., P.T. Chandler, E.B. Kalan, A.R. Loeblich III, G. Fuller, and A.A. Benson. 1967. Food value of a red tide (Gonyaulax polyedra). Science 158, 789-790.

Prakash, A. 1963. Source of paralytic shellfish toxin in the Bay of Fundy. J. Fish. Res. Bd. Can. 20, 983-996.

Prakash, A., and F.J.R. Taylor. 1966. A "red water" bloom of Gonyaulax acatenella in the Strait of Georgia and its relation to paralytic shellfish toxicity. J. Fish. Res. Board Can. 23, 1265-1270.

Rapoport, H., M.S. Brown, R. Oesterlin, and W. Schuett. 1964. Saxitoxin. 147th National Meeting, American Chemical Society, Philadelphia, Penna.

Ray, S.M., and D.V. Aldrich. 1965. Gymnodinium breve: Induction of shellfish poisoning in chicks. Science 148, 1748-1749.

Ray, S.M., and D.V. Aldrich. 1967. Ecological interactions of toxic dinoflagellates and molluscs in the Gulf of Mexico. Animal Toxins 75-83. Editors, F.E. Russel and P.R. Saunders, Pergamon Press, Oxford and New York 1967.

Reich, K., F. Bergmann, and M. Kidron. 1965. Studies on the homo-
     geneity of Prymnesin, the toxin isolated from Prymnesium
     parvum Carter. Toxicon 3, 33-39.
Robinson, G.A. 1968. Distribution of Gonyaulax tamarensis Lebour
     in the Western North Sea in April, May and June 1968. Nature
     220, 22-23.
Sasner, J.J., Jr. 1965. A study of the effects of a toxin produced
     by the Florida red tide dinoflagellate Gymnodinium breve Davis.
     Ph.D. Thesis. University of California, Los Angeles,
     California.
Sawyer, P.J., J.H. Gentile, and J.J. Sasner, Jr. 1968. Demonstra-
     tion of a toxin from Aphanizomenon flos-aquae (L.) Ralfs.
     Can. J. Microbiol. 14, 1199-1204.
Schantz, E.J., J.D. Mold, D.W. Stanger, J. Shavel, F.J. Riel, J.P.
     Bowden, J.M. Lynch, R.W. Wyler, B. Riegel, and H. Sommer.
     1957. Paralytic shellfish poison. VI. A procedure for the
     isolation and purification of the poison from toxic clam and
     mussel tissue. J. Am. Chem. Soc. 79, 5230.
Schantz, E.J., J.D. Mold, W.L. Howard, J.P. Bowden, D.W. Stanger,
     J.M. Lynch, O.P. Wintersteiner, J.D. Dutcher, D.R. Walters,
     and B. Riegel. 1961. Paralytic shellfish poison. VIII.
     Some chemical and physical properties of purified clam and
     mussel poisons. Can. J. Chem. 39, 2117-2123.
Schantz, E.J., and H.W. Magnusson. 1964. Observations on the
     origin of the paralytic poison in Alaska butter clam. J.
     Protozool. 11, 239-242.
Schantz, E.J., J.M. Lynch, G. Vayvada, K. Matsumoto, and H.
     Rapoport. 1966. The purification and characterization of
     the poison produced by Gonyaulax catenella in axenic culture.
     Biochemistry 5, 1191-1195.
Schantz, E.J. 1968. Biochemical studies on certain algal toxins.
     Biochemistry of Some Foodborne Microbial Toxins 51-65.
     Editors, R.I. Mateles, and G.N. Wogan, MIT Press, Cambridge,
     Massachusetts 1968.
Schradie, J., and C.A. Bliss. 1962. The cultivation and toxicity
     of Gonyaulax polyedra. Lloydia 25, 214-221.
Shilo, M., and R.F. Rosenberger. 1960. Studies on the toxic
     principles formed by the chrysomonad Pyrmnesium parvum Carter.
     Ann. N.Y. Acad. Sci. 90, 866-876.
Shilo, M. 1967. Formation and mode of action of algal toxins.
     Bacteriological Reviews 180-193.
Sommer, H., and K.F. Meyer. 1937. Paralytic shellfish poisoning.
     A.M.A. Arch. Pathol. 24, 560-598.
Sommer, H., W.F. Whedon, C.A. Kofoid, and R. Stohler. 1937.
     Relation of paralytic shellfish poison to certain plankton
     organisms of the genus Gonyaulax. A.M.A. Arch. Pathol. 24,
     537-559.
Starr, T.J. 1958. Toxin from Gymnodinium brevis. Texas Repts.
     Biol. and Med. 16, 500-507. From 1960 Chemical Abstract.
     vol. 54.

Thompson, L.  To The American Indian 28, Cummins Print Shop.
    Eureka, California 1916.
Ulitzur, S., and M. Shilo. 1966.  Mode of action of Prymnesium
    parvum ichthyotoxin.  J. Protozool. 13, 332-336.
Wood, P.D. 1968.  Dinoflagellate crop in the North Sea. Nature
    220, 21.

# KINETICS OF ALGAL GROWTH IN AUSTERE MEDIA

G. C. McDonald[1], R. D. Spear, P. J. Lavin[2], N. L. Clesceri

Rensselaer Polytechnic Institute, Troy, New York

Man the builder has become man the destroyer. In his ever expanding quest for progress and power he has succeeded in adversely affecting his environment. One of the paramount effects of man's despoliation of the environment has been the accelerated ageing of natural bodies of water. This eutrophication has evidenced itself in the proliferation of nuisance algal growths. Previous investigations have indicated that nitrogen and phosphorus can be the primary cause of this imbalance in the aquatic ecosystem.

## Materials and Methods

The object of the present investigation was to ascertain the effect of various concentrations of nitrogen and phosphorus on the growth rate of a test algal organism, namely Selenastrum capricornutum a uni-cellular green alga. This organism was selected because of its ease of culturing in the laboratory and for the fact that contrary to algae usually used in the laboratory, for example Chlorella, this organism has produced nuisance blooms in lake systems notably in Europe. The media selected for this study were the Basic ASM medium of the provisional algal assay procedure (PAAP, 1969) and a modification of Gorham's medium (Hughes et al., 1958). These media were selected to insure minimal nutrient carry over and to approximate as closely as possible the natural environment. The modification noted for Gorham's medium involved a 10 to 1 dilution of the standard medium with an increase in the sodium carbonate

---

[1] Albany County Sewer District, Albany, New York
[2] Albany County Health Department, Albany, New York

concentration to 50 milligrams per liter as an aid to pH control.
Tables 1 and 2 present the media employed in the study. Prior to
testing at the various nutrient concentrations, the organism was
maintained in Basic ASM and modified tenth Gorham's media. Experiments were conducted at a constant illumination of 550 foot candles
and a temperature of 23 ± 1°C (Skulberg, 1964; Skulberg, 1966).

The culture method was the use of bubbling tubes, that is 25
by 200 mm optically matched pyrex test tubes filled with 25 mm of
growth media which is continuously aerated and mixed by passing
water saturated air through a section of glass tubing that is ver-
tically placed to the bottom of the culture tube. The growth of
the cultures was followed using a Bausch and Lomb Spectronic 20 at
750 mm and a one inch cell path (Yentsch, 1957).


                            Experimental Results

The experimentation consisted in obtaining growth rates of the
test alga in Basic ASM and modified tenth Gorham's media, both of
which used glass distilled water and/or 0.45μ membrane-filtered
Lake George water (an oligotrophic soft water lake in the eastern
Adirondack Mountains) as dilution water. The rationale for using
glass distilled water and Lake George water was to allow inter-
pretation of the adequacy of the synthetic media for alga culturing.
Once these growth rates were established, the concentrations of
nitrogen and phosphorus in the two media were fractionally reduced
to one half and one quarter of the full amount and the growth rate
at each concentration level ascertained.

For the following illustrations, the ordinate indicates the
extent of growth attained in 24 hours on a relative basis with
Basic ASM medium made up with glass distilled water the norm for
comparison since this medium supported the greatest rate of growth.
During the experimentation only the component under study was al-
tered from its concentration in the original medium. As illustrated in Figure 1, the effect of nitrogen reductions on the growth
rate is less severe than fractional reductions in the phosphorus
concentrations. The growth rate for Selenastrum capricornutum in
Basic ASM is 1.32. In noting the effect of nitrogen reduction, it
can be seen that a 50% reduction in nitrogen concentration produced a growth rate of 0.96 or a one third reduction in cell concentration over the 24 hour period. This is contrasted with a
reduction in K rate to 0.74 when the nitrogen is further reduced
from .5 nitrogen to 0.25 nitrogen. The overall change in cell
concentration produced by a decrease in nitrogen concentration
from the full nitrogen level to one fourth was 40%.

The influence of changes in phosphorus concentration is also
illustrated in Figure 1. A reduction in cell concentration of

| Compound | Stock Solution (g/l) | For 1 Liter Of Solution Use (ml) | Final Concentration In Medium (mg/l) |
|---|---|---|---|
| $NaNO_3$ | 8.50 | 10 | 85.0 |
| $K_2HPO_4$ | 0.348 | 10 | 3.48 |
| $CaCl_2 \cdot 2H_2O$ | 1.47 | 10 | 14.7 |
| $Na_2CO_3$ | 5.00 | 10 | 50.0 |
| $MgSO_4 \cdot 7H_2O$ | 4.90 | 10 | 49.0 |
| $MgCl_2 \cdot 6H_2O$ | 1.90 | 10 | 19.0 |
| $FeCl_3$ | 0.032 | 10 | 0.32 |
| $Na_2 \cdot EDTA$ | 0.100 | | 1.0 |

Glass Distilled Or Lake George Water – Dilute To 1 Liter.

Table 1.   Basic ASM Medium

| Compound | Stock Solution (g/l) | For 1 Liter Of Solution Use (ml) | Final Concentration In Medium (mg/l) |
|---|---|---|---|
| $NaNO_3$ | 49.60 | 1 | 49.6 |
| $K_2HPO_4$ | 3.90 | 1 | 3.9 |
| $CaCl_2 \cdot 2H_2O$ | 3.60 | 1 | 3.6 |
| $Na_2CO_3$ | 5.00 | 10 | 50.0 |
| $MgSO_4 \cdot 7H_2O$ | 7.50 | 1 | 7.5 |
| $NaSiO_3 \cdot 9H_2O$ | 5.80 | 1 | 5.8 |
| $FeC_6H_5O_7 \cdot 3H_2O$ | 0.60 | 1 | 0.6 |
| $H_3C_6H_5O_7 \cdot H_2O$ | 0.60 | | 0.6 |
| $Na_2 \cdot EDTA$ | 0.10 | | 0.1 |

Glass Distilled Or Lake George Water – Dilute To 1 Liter.

Table 2.   Modified 1/10 Gorham's Medium

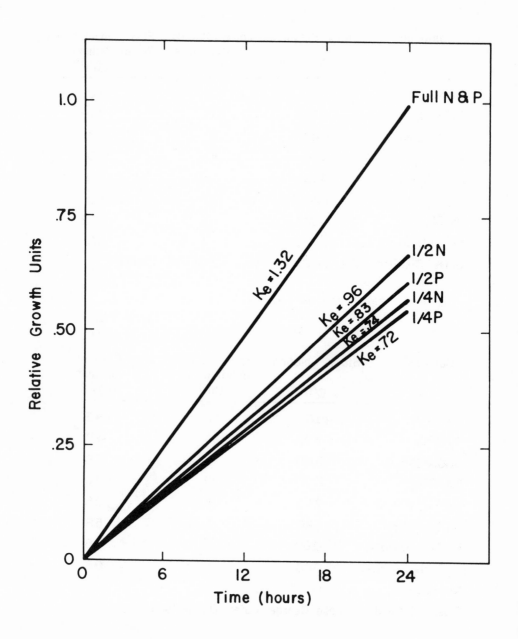

Fig. 1.   Effect of Varying Nutrient Concentrations in ASM
          (DD) For <u>Selenastrum</u> <u>capricornutum</u>.

40% was affected by a 50% reduction in phosphorus level. A further
reduction in the phosphorus level to one fourth of the full medium
phosphorus level produced an overall reduction in cell concentration of 45%. The experiments indicated in Figure 1 and the following figures are illustrative of the nutrient limitation exhibited
by the decreased concentrations of the required elements, nitrogen
and phosphorus, with the organism being consistently more sensitive
to reductions in phosphorus concentration than nitrogen concentration.

In Figure 2 the data were obtained by using the Basic ASM
medium with Lake George water substituted for glass distilled water.
The striking result evident in these data is the substantial reduction in growth rate for the full complement of nitrogen and
phosphorus from 1.32 with glass distilled water as diluent to a
value of 0.98 with Lake George water as diluent. The substitution
of Lake George water as diluent for the Basic ASM medium thus
produces a 29% reduction in cell mass production. A similar effect was noted in the growth rate of Selenastrum capricornutum in
modified tenth Gorham's medium due to the Lake George water diluent.
It appears that Lake George water contains some factor inhibitory
to the growth of Selenastrum capricornutum with greater inhibition
prevalent in the Basic ASM medium. The growth rates produced by
fractional reductions in phosphorus are comparable to those obtained with these same reductions with glass distilled water.
However, the effect of a reduction in nitrogen concentration to
the one half level was more pronounced with Lake George water as
diluent than with glass distilled water as diluent. The K rate
for the one half nitrogen level in glass distilled water was 0.96,
whereas the K value for Lake George water was 0.72, a 20% decrease
in cell concentration due to Lake George water.

The basal medium for experimentation illustrated in Figure 3
was modified tenth Gorham's medium with glass distilled water as
diluent. It can be seen that the growth rate for the full complement of nutrients is 1.15 which is considerably higher than the
value obtained for Basic ASM with Lake George water and modified
tenth Gorham's with Lake George water. The difference noted in K
rate for a reduction of nitrogen to the one half level that is,
a value of 1.03 and a concomitant 25% reduction in cell concentration, was not as pronounced as the reductions noted for a
similar decrease in nitrogen level with Basic ASM, using either
glass distilled water or Lake George water. When the nitrogen was
fractionally reduced to one fifth its original level there was
drastic reduction in the growth rate to a value of 0.40 and a 50%
reduction in cell concentration. This indicated that for unimpeded
growth a value at least one half of the full nitrogen concentration is required. For a reduction in phosphorus concentration to
one half the level originally posited a growth rate of 0.60 was
obtained as compared to 1.15 for the full complement of phosphorus.

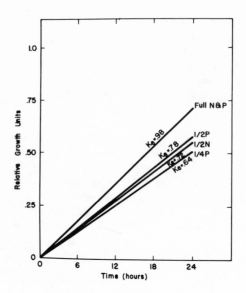

Fig. 2.  Effect of Varying Nutrient Concentrations in
ASM (LG) For Selenastrum capricornutum.

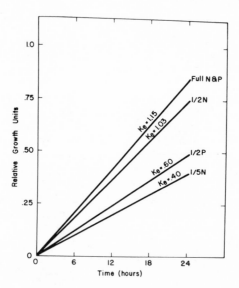

Fig. 3.  Effect of Varying Nutrient Concentrations in
Modified .1 G (DD) For Selenastrum capricornutum.

This reduction in growth rate represents a decrease of 40% in cell concentration. As noted previously, a reduction to one half phosphorus level in Basic ASM also produces a 40% change in cell concentration. A predictable result since both media have approximately the same phosphorus concentration and the organism is phosphorus limited in both situations.

Based upon the data presented in Figures 1 and 3 and the relative concentrations of nitrogen and phosphorus in both media, a comparison can be made between the results of Basic ASM with one half nitrogen and modified tenth Gorham's medium. The Basic ASM medium with one half nitrogen containing 7 mg/l and the tenth Gorham's medium containing 8 mg/l and approximately the same phosphorus concentration in both media. The K rate obtained for the modified tenth Gorham's medium is 1.15 while the K rate for Basic ASM with one half nitrogen is 0.96. For a 15% increase in nitrogen level one obtains 20% greater cell concentration modified tenth Gorham's medium as opposed to the Basic ASM medium with one half nitrogen. These proportions again hold when one compares Basic ASM with one fourth nitrogen and modified tenth Gorham's with one half nitrogen. The respective growth rates being 0.74 for the former and 1.03 for the latter, again a 20% increase in cell concentration for a 15% increase in nitrogen level. As depicted in Figure 4 higher growth rates were obtained with glass distilled water as diluent than with Lake George water as diluent in both Basic ASM medium and modified tenth Gorham's medium. All curves indicate a predictable effect on the growth rate with increasing concentrations of phosphorus, that is, higher values of phosphorus yield higher growth rates. In all cases within the range of concentrations of phosphorus indicated, phosphorus is limiting. However, the limitation is less pronounced with Basic ASM using Lake George water as diluent. It appears that the alga cannot utilize the increased phosphorus concentration because of some inhibitory factor present in Lake George water.

The data presented in Figure 5 are a summation of experimentations with various nitrogen concentrations in modified tenth Gorham's medium with glass distilled water as diluent and Basic ASM medium with either glass distilled water or Lake George water. The curve for modified tenth Gorham's medium with glass distilled water indicates that nitrogen is present in excess in the full medium. If interest were placed on the minimization of nutrient storage and carry over, the nitrogen level could be reduced in this medium to one half the level presently prescribed. For Basic ASM medium once again the growth rates were higher with glass distilled water than with Lake George water. Increasing concentrations of nitrogen elicited higher growth rates and within the range of concentrations indicated nitrogen was limiting for Basic ASM medium.

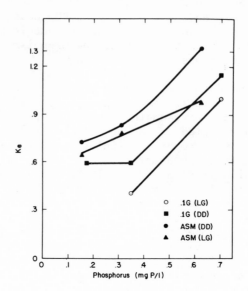

Fig. 4.   Effect of Varying Phosphorus Concentrations On
Values of $K_e$ For <u>Selenastrum</u> <u>capricornutum</u>.

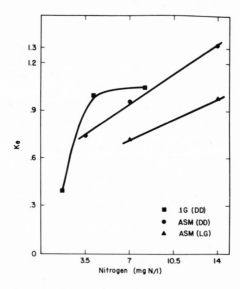

Fig. 5.   Effect of Varying Nitrogen Concentrations On
Values of $K_e$ For <u>Selenastrum</u> <u>capricornutum</u>.

In summary it can be stated that there appears to be an inhibition to the growth of the organism as indicated by reduced growth rates when Lake George water is used as a diluent for the basic components of either modified tenth Gorham's medium or Basic ASM medium.  To this point, further particular investigation is now being carried on, to ascertain the extent and the causative factor of such inhibition.  It was also demonstrated that the concentration of nitrogen in modified tenth Gorham's medium may be reduced to one half the posited level without any significant change in the growth rate of <u>Selenastrum capricornutum</u>.

Note: Phosphorous was always tested in the form of phosphate in all this work.

## References

Joint Industry/Government Task Force on Eutrophication, 1969. <u>Provisional Algal Assay Procedure</u>.

Hughes, E.O., Gorham, P. and Zehnder, A., 1958.  "Toxicity of Unialgal Culture of Microcystis aeruginosa", Canadian Journal of Microbiology, <u>4</u>, 225-236.

Skulberg, O.M., 1964.  "Algae Problems Related to the Eutrophication of European Water Supplies and a Bio-Assay Method to Assess Fertilizing Influence of Pollution on Inland Waters", <u>Algae and Man</u>, D.F. Jackson, ed., Plenum Press, New York, New York, 262-299.

Skulberg, O.M., 1966.  "Algal Cultures as a Means to Assess the Fertilizing Influence of Pollution", Journal Water Poll. Control Fed., <u>38</u>, 319-320.

Yentsch, C.S., 1957.  "A Non-Extractive Method for the Quantitative Estimation of Chlorophyll in Algal Cultures", Nature, <u>179</u>, 1302-1304.

# HETEROTROPHIC NUTRITION OF WASTE-STABILIZATION POND ALGAE

Varley E. Wiedeman

Department of Biology, University of Louisville

Louisville, Kentucky  40208

Waste-stabilization ponds are man-made eutrophic aquatic habitats into which man-made wastes - household to industrial - are placed to be incorporated into "stabilized" living organic matter. The living system of most waste-stabilization ponds consists in the greater part of bacteria, fungi, and algae, usually considered to function as in Figure 1. Although the algae are considered to be photosynthetic, they can only accomplish this process during daylight hours and, due to density of materials in the ponds themselves, only within the top foot or so from the surface. During hours of darkness and up to several feet from the bottom of the pond algae become oxygen requiring and carbon dioxide releasing organisms, in this respect functioning in the same manner as the bacteria and fungi. Samples from waste-stabilization ponds reveal that living algae are nearly as numerous in the aphotic zone as in the photic. If they are not photosynthetically incorporating carbon dioxide for their growth their metabolic functions may further parallel those of the saprobic bacteria and fungi.

Numerous reports of associations of algae with waste treatment have been made. A selected listing of some of these are give in Table 1.

As a part of a larger study concerned with algal ecology in a waste-stabilization pond system (Wiedeman 1964, 1965) axenic isolates representing 11 genera of green algae were examined for their ability to utilize several carbon sources (glucose, mannose, fructose, urea, sodium acetate, casein hydrolysate, and coconut milk) under a variety of environmental conditions. The algae were isolated from the Austin, Texas waste-stabilization ponds using modifications of the techniques utilized by Pringsheim (1946)

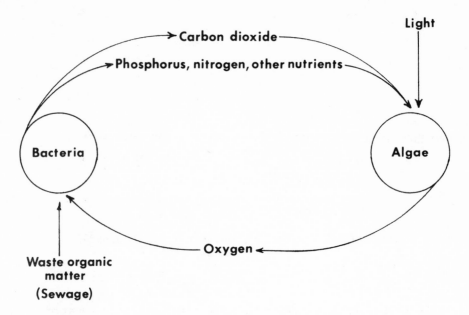

Fig. 1.  Bacteria and algae cycle of a waste-stabilization pond
(After Sidio *et al.*, 1961)

Table 1

Studies concerned with algae in waste treatment

| | |
|---|---|
| Chick (1903) | Bartsch (1960) |
| Purdy (1937) | Maloney and Robinson (1961) |
| Giesecke and Zeller (1939) | Bartsch (1961) |
| Myers (1948) | Eppley and MaciasR (1962, 1963) |
| Ludwig *et al.* (1951) | Ganapati and Bopardikar (1962) |
| Ludwig and Oswald (1952) | Porges and Mackenthun (1963) |
| Oswald *et al.* (1953) | Wiedeman (1964, 1965) |
| Silva and Papenfuss (1953) | Wiedeman and Bold (1965) |
| Allen (1955) | Palmer (1967) |
| Maloney (1959) | Owens and Wood (1968) |
| Pipes and Gotaas (1960) | |

and Brown and Bischoff (1962) by atomizing (Wiedeman, Walne, and
Trainor, 1964) washed algal cultures onto the solidified surface
of tris-buffered inorganic medium, TBIM (Smith and Wiedeman, 1964).
The media used for the tests of heterotrophic nutrition of the
algal isolates are listed in Table 2.

The respective carbon sources were sterile filtered into autoclaved TBIM containing 1.5 percent agar the whole being aseptically dispensed into sterile culture tubes. The environmental conditions to which the tests were subjected are listed in Table 3.

Table 2

Media used in tests for heterotrophic growth

Tris-buffered Inorganic Medium (TBIM) (Control)
TBIM + 0.01 M glucose
TBIM + 0.01 M mannose
TBIM + 0.01 M fructose
TBIM + 0.01 M urea
TBIM + 0.01 M sodium acetate
TBIM + 0.6 g/liter casein hydrolysate
TBIM (90%) + coconut milk (10%)

Table 3

Environmental conditions to which the
algal cultures were submitted

12 hours light - 12 hours dark photoperiod - aerobic
12 hours light - 12 hours dark photoperiod - anaerobic*
Total darkness - aerobic
Total darkness - anaerobic*

All at 22 C.

* 1.5 ml sterile mineral oil over inoculated agar to
  exclude exogenous air

Six weeks after inoculation the cultures were examined for growth. On the basis of the results (Tables 4 and 5) several heterotrophic categories can be discerned. Those organisms capable of growth only by photosynthetically incorporating carbon dioxide (Category A) may be called, as is traditional, *obligate photoautotrophs*. Those algae capable of utilizing a preformed organic carbon source only in the light when exogenous air is

excluded (Category B) are *photoheterotrophs*. The algae that incorporate preformed carbon sources in the light with exogenous air
excluded or in the dark with air present (Category C) may function
as *photo- and/or aerobic heterotrophs*. Finally, those algae
capable of incorporation of the carbon sources under all conditions
are considered as *euheterotrophs* (Category D).

As a result of this variety of heterotrophic possibilities the
algae should be included with the bacteria and fungi as being agle
to incorporate organic materials from their environment. The
diagramatic representation of a waste-stabilization pond, taking
these facts into consideration, would then be as in Figure 2.

In consideration of algae in relation to waste treatment there
are several additional areas which should be investigated: enzyme
production, protein content, mineral nutrition etc.

Other algal groups, particularly members of the Divisions
Cyanophycophyta and Euglenophycophyta, should be isolated into
axenic cultures from waste-stabilization ponds and subjected to
studies similar to those described above.

In view of the differences exhibited by the algae in incorporation of the several carbon sources tracer studies would aid in
evaluation of metabolic pathways involved and contribute to the
understanding of other metabolic processes in the algae.

Equipment more refined than culture tubes with mineral oil,
namely fermentation units and gas incubators equipped with light
sources for photosynthetic studies, would insure more critical
evaluations of the heterotrophic processes.

---

Table 4

Environmental conditions which supported (+) or
did not support (-) growth of the algal isolates

|          | Light-Dark |           | Dark Only |           |
|----------|:----------:|:---------:|:---------:|:---------:|
| Category | Aerobic    | Anaerobic | Aerobic   | Anaerobic |
| A        | +          | −         | −         | −         |
| B        | +          | +         | −         | −         |
| C        | +          | +         | +         | −         |
| D        | +          | +         | +         | +         |
| I        | I N H I B I T I O N |    |           |           |

Table 5

Numbers of isolates of the algal genera which exhibited
growth in the various autotrophic and heterotrophic categories

| Genus | Total number isolates | TBIM A | B | C | D | I | Glucose A | B | C | D | I | Mannose A | B | C | D | I | Fructose A | B | C | D | I |
|---|---|---|---|---|---|---|---|---|---|---|---|---|---|---|---|---|---|---|---|---|---|
| *Chlamydomonas* | 1 | 1 | | | | | | | | | 1 | 1 | | | | | 1 | | | | |
| *Chlorococcum* | 5 | 5 | | | | | 3 | | | 2 | | 3 | 1 | 1 | 1 | | 5 | | | | |
| *Dictyosphaerium* | 2 | 2 | | | | | 1 | 1 | | | | 2 | | | | | 2 | | | | |
| *Pediastrum* | 2 | 2 | | | | | | | | 2 | | | | | | | 2 | | | | |
| *Coelastrum* | 5 | 5 | | | | | | 1 | 1 | 3 | | 2 | 1 | 1 | 3 | | 5 | | | | |
| *Chlorella* | 6 | 6 | | | 1 | | 1 | 1 | 1 | 4 | 1 | 2 | 1 | | 4 | | | 3 | | | 3 |
| *Ankistrodesmus* | 3 | 3 | | | | | | 1 | 1 | 1 | | 1 | 1 | | 1 | | 1 | 2 | | | |
| *Dactylococcus* | 2 | 2 | | | | | | 1 | | 1 | | | | 2 | | | 1 | 1 | | | |
| *Scenedesmus* | 11 | 11 | | | | | 2 | | 6 | 3 | | | 3 | 7 | 1 | | 9 | 2 | | | |
| *Chlorosarcinopsis* | 1 | 1 | | | | | 1 | | | | | 1 | | | | | 1 | | | | |
| *Stigeoclonium* | 1 | 1 | | | | | 1 | | | | | 1 | | | | | 1 | | | | |

| Genus | Total number isolates | Urea A | B | C | D | I | Na acetate A | B | C | D | I | Cas hydrol A | B | C | D | I | Cocon milk A | B | C | D | I |
|---|---|---|---|---|---|---|---|---|---|---|---|---|---|---|---|---|---|---|---|---|---|
| *Chlamydomonas* | 1 | 1 | | | | | 1 | | | | | 1 | | | | | 1 | | | | |
| *Chlorococcum* | 5 | 5 | | | | | 5 | | | | | 5 | | | | | 4 | 1 | | | |
| *Dictyosphaerium* | 2 | 2 | | | | | 1 | 1 | | | | 2 | | | | | | | | | 1 |
| *Pediastrum* | 2 | 2 | | | | | 1 | 1 | | | | 2 | | | | | | | | | 2 |
| *Coelastrum* | 5 | 5 | | | | | 3 | 2 | | | | 5 | | | | | 4 | | | | 1 |
| *Chlorella* | 6 | 5 | | | 1 | | | 6 | | | | 2 | 3 | | 1 | | | 3 | | | 3 |
| *Ankistrodesmus* | 3 | 3 | | | | | 3 | | | | | 1 | 2 | | | | | 2 | | | 1 |
| *Dactylococcus* | 2 | 2 | | | | | 2 | | | | | 2 | | | | | | 2 | | | |
| *Scenedesmus* | 11 | 11 | | | | | 6 | 5 | | | | 7 | 4 | | | | | 7 | | | 3 |
| *Chlorosarcinopsis* | 1 | 1 | | | | | 1 | | | | | 1 | | | | | 1 | | | | |
| *Stigeoclonium* | 1 | 1 | | | | | | | | | | Cultures contaminated | | | | | | | | | |

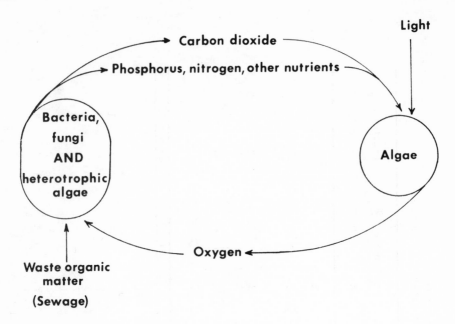

Fig. 2.  Suggested bacteria, fungi, and algae cycle of a waste-
        stabilization pond system


    Consideration should be given as to whether some algal growth
is supported by organic gases (methane, etc.) known to be produced
by sewage treatment processes.

SUMMARY

    Waste-stabilization ponds are a rich source of organic and
inorganic nutrients for the growth of three plant groups - bacteria,
fungi, and algae.  Bacterial and fungal representatives are known to
be capable of digesting and assimilating organic carbon sources
under aerobic or anaerobic conditions in either light or darkness.
Algae are generally thought of as assimilating carbon for growth as
a result of photosynthetic carbon dioxide fixation.  Abundant algae
appear in the aphotic zone of waste-stabilization ponds (which may
be anaerobic in part) suggesting that some algae are capable of
metabolic processes similar to those of bacteria and fungi.
Experimentation with axenic green algal cultures indicates that such
is the case; certain algae are capable of assimilating preformed
carbon sources under all conditions and are considered to be
euheterotrophs.  Some are capable of assimilating the preformed
carbon sources in the light whether aerobic or anaerobic and are

considered to be photo- and/or aerobic heterotrophs. Some are able to assimilate the preformed carbon sources only in the light whether aerobic or anaerobic and are considered to be photohetero-trophs. Some algae can not assimilate the preformed carbon sources under any conditions and are considered to be the obligate photo-autotrophs. Finally, some algae are inhibited by the preformed carbon sources.

## LITERATURE CITED

Allen, M. B. 1955. General features of algal growth in sewage oxidation ponds. State Water Pollution Control Board, Sacramento, California.

Bartsch, A. F. 1960. Algae in relation to oxidation processes in natural waters. In: The Pymatuning Symposia in Ecology, The Ecology of Algae, Ed. by C. A. Tryon, Jr. and R. T. Hartman, Special Publication Number 2, Pymatuning Laboratory of Field Biology, University of Pittsburgh, Pittsburgh, Pennsylvania.

————. 1961. Algae as a source of oxygen in waste treatment. J. Water Pollution Control Federation. 33, 239-249.

Brown, R. M., Jr., and H. W. Bischoff. 1962. A new and useful method for obtaining axenic cultures of algae. Phycol. News Bull. 15, 43-44.

Chick, H. 1903. A study of a unicellular green alga, occurring in polluted water, with especial reference to its nitrogenous metabolism. Proc. Royal Soc. London. 71, 458-476.

Eppley, R. W. and F. M. MaciasR. 1962. Rapid growth of sewage lagoon Chlamydomonas with acetate. Physiol. Plant. 15, 72-79.

————, and ————. 1963. Role of the alga Chlamydomonas mundana in anaerobic waste stabilization lagoons. Limnol. and Oceanogr. 8, 411-416.

Ganapati, S. V. and M. V. Bopardikar. 1962. A comparative study of natural purification of flowing sewage in the sewage farms at Ahmedabad and Bangalore. J. Inst. of Engineers (India). 42, 672-689.

Giesecke, F. E., and P. J. A. Zeller. 1939. Treatment of settled sewage in lakes. Bull. of the A. and M. College of Texas, 4th Series, Vol. 10, No. 47.

Ludwig, H. F., W. J. Oswald, H. B. Gotaas, and V. Lynch. 1951. Algae Symbiosis in Oxidation Ponds. I. Growth characteristics of Euglena gracilis cultured in sewage. Sewage and Ind. Wastes. 23, 1337-1355.

————, and W. J. Oswald. 1952. Role of algae in sewage oxidation ponds. Sci. Monthly. 74, 3-6.

Maloney, T. E. 1959. Utilization of sugars in spent sulfite liquor by a green alga, Chlorococcum macrostigmatum. Sewage and Ind. Wastes. 31, 1395-1400.

_____, and E. L. Robinson. 1961. Growth and respiration of a green alga in spent sulfite liquor. Tech. Assoc. of the Pulp and Paper Ind. 44, 137-141.

Myers, J. 1948. Studies of sewage lagoons. Pub. Works. 79, 25-27.

Oswald, W. J., H. B. Gotaas, H. F. Ludwig, and V. Lynch. 1953. Algae symbiosis in oxidation ponds. III. Photosynthetic oxygenation. Sewage and Ind. Wastes. 25, 692-705.

Owens, M. and G. Wood. 1968. Some aspects of the eutrophication of water. Water Research. 2, 151-159.

Palmer, C. M. 1967. Nutrient assimilation by algae in waste stabilization ponds. Proc. Indiana Acad. Sci. 76, 204-209.

Pipes, W. O. and H. B. Gotaas. 1960. Utilization of organic matter by *Chlorella* grown in sewage. Appl. Microbiol. 8, 163-169.

Porges, R. and K. M. Mackenthun. 1963. Waste stabilization ponds: Use, function, and biota. Biotech. and Bioenginer. 5, 255-273.

Pringsheim, E. G. 1946. Pure Cultures of Algae. Cambridge Univ. Press.

Purdy, W. C. 1937. Experimental studies of natural purification in polluted waters. X. Reoxygenation of polluted waters by microscopic algae. Pub. Health Reports. 52, 945-978.

Sidio, A. D., R. T. Hartman, and P. Fugazzotto. 1961. First domestic waste stabilization pond in Pennsylvania. Public Health Reports. 76, 201-208.

Silva, P. C., and G. F. Papenfuss. 1953. A systematic study of the algae of sewage oxidation ponds. State Water Pollution Control Board, Sacramento, California.

Smith, R. L. and V. E. Wiedeman. 1964. A new alkaline growth medium for algae. Can. J. Bot. 42, 1582-1586.

Wiedeman, V. E. 1964. Some aspects of algal ecology in a waste-stabilization pond system. *Diss.*, Univ. of Texas.

_____. 1965. Chemical analyses and algal composition in a waste-stabilization pond system. Southwestern Nat. 10, 188-210.

_____, and H. C. Bold. 1965. Heterotrophic growth of selected waste-stabilization pond algae. J. Phycol. 1, 66-69.

_____, P. L. Walne, and F. R. Trainor. 1964. A new technique for obtaining axenic cultures of algae. Can. J. Bot. 42, 958-959.

# FATTY ACIDS OF BLUE-GREEN ALGAE

Raymond W. Holton and Harry H. Blecker

Department of Botany, University of Tennessee, Knoxville.

Tennessee 37916 and Department of Chemistry, University

of Michigan Flint College, Flint, Michigan 48503

The Cyanophyta, or blue-green algae, is a small group of primitive organisms. The cells of these algae are very simple in their
structure and like the bacteria, they are said to be prokaryotic in
contrast to more complex eukaryotic cellular organization found in
other plants and in animals. In prokaryotic cells, the limiting
membranes and therefore the discrete organelles such as the mito-
chondria, chloroplasts, and nuclei that are found in cells of
eukaryotic organisms, are absent. The biochemical processes associated with these structures, i.e., respiration, photosynthesis,
and DNA replication and RNA transcription, nevertheless go on in
prokaryotic cells. The prokaryotic cell type is considered to be
a very primitive evolutionary condition and this primitiveness has
been confirmed by findings in recent years of fossils which look
very much like present day Cyanophyta and which have been dated as
being two to three billion years old. Thus there is a good evidence that the blue-green algae evolved very early in biological
evolution and in fact, it has been suggested that these organisms
were the first oxygen-evolving photosynthetic cells on the earth.

In the natural environment, blue-green algae are found in a
variety of habitats. While they are commonly found on soil and in
water, they often are seen on and can be isolated from the barks
of trees and from lichens in which are associated symbiotically
with fungi. Other associations occur in the water fern, _Azolla_
and in cycads. Thus though the analyses which are presented below
are from algal cultures grown in various laboratories, it should
be borne in mind that, in nature, these algae are found in diverse

habitats ranging from the high temperatures of hot springs to the
low temperatures of the Antarctic soil and water.

It might be assumed a priori, that the fatty acid content of
a small group of organisms such as the Cyanophyta would be quite
uniform.  On the other hand, because they probably evolved a long
time ago, there has been sufficient time to permit the evolution
of different biosynthetic pathways if an exact fatty acid composition is not critical to the cell's structure and/or function.
Likewise, if the variety of habitats in which these organisms are
found influence their chemical composition, then we might expect
to find variations in the fatty acid patterns from one species to
another.

The first published analysis of the fatty acid composition
of a blue-green alga was done by Mazur and Clarke (1942).  The
complexity of the chemical techniques necessary for fatty acid
analyses at the time is suggested by the fact that these investigators started with 289 grams dry weight of Gloeotrichia as
opposed to the 1 gram or less of sample of algae now used in gas
chromatographic procedures and that in their analyses, they were
able only to divide the isolated acids into a saturated or an unsaturated group.  Gas chromatography was not put to use on the blue-
green algae until 1964 when two different species were analyzed in
two different laboratories.  Levin et al. (1964) examined Anabaena
variabilis and found mono-, di-, and triunsaturated acids including
α-linolenic acid, a triunsaturated acid found in all green tissues
which were known at that time to carry out green plant photosynthesis.  These authors suggested that the linolenic acid may be
somehow involved in the oxygen evolution part of photosynthesis
since photosynthetic bacteria which do not evolve oxygen lack this
acid.  Our own work with Anacystis nidulans (Holton et al., 1964)
showed a quite different pattern as the fatty acid composition was
very simple and the only unsaturated acids present were the mono-
unsaturated ones.  As is shown in Table I, the acids present were
principally $C_{16}$ acids, the saturated, palmitic, and the monounsat-
urated, palmitoleic.  Under all conditions of growth in which the
temperature was varied and other conditions kept constant, used in
our study, these two fatty acids made up at least 85% of the total
in the alga.  In these experiments the growth temperature clearly
affected the quantitative composition of the acids but no significant qualitative differences were noted.

Other investigators have also examined Anacystis, a favorite
laboratory organism for many kinds of physiological and biochemical
experiments, and the similarities of the fatty acid pattern obtained
in different laboratories is striking.  It should be noted that in
the data of Oró et al. (1967) gas chromatography was combined with
mass spectrometry and various isomers were detected that other
investigators have not detected and also that 18% of the acid

## TABLE I

### FATTY ACID COMPOSITION OF ANACYSTIS NIDULANS IN DIFFERENT EXPERIMENTS

| Algal Growth Temp. °C. | Percentages (by weight of total) of Various Fatty Acids Carbon atoms:double bonds | | | | | | | | | | | % Lipid of Algal Dry wt. | Ref. |
| --- | --- | --- | --- | --- | --- | --- | --- | --- | --- | --- | --- | --- | --- |
| | 12:0 | 14:0 | 14:1 | 15:1 | 16:0 | 16:1 | 17:1 | 18:0 | 18:1 | 20:0 | Others | | |
| 26 | tr. | 0.9 | 3.3 | | 47.3 | 44.1 | 0.6 | 0.6 | 2.7 | 0.5 | | 11 | a |
| 32 | tr. | 0.5 | 1.0 | | 47.0 | 38.8 | 0.5 | 1.4 | 10.0 | 0.7 | | | a |
| 35 | tr. | 0.6 | 1.3 | | 47.3 | 40.2 | 0.8 | 0.8 | 8.5 | 0.7 | | | a |
| 41 | | 0.8 | 0.7 | | 57.6 | 31.3 | 0.5 | 0.5 | 8.0 | 0.6 | | | a |
| 39 | tr. | 0.5 | 0.6 | | 49. | 34. | | 1.7 | 13. | | 15:0 = tr. | | b |
| | | | 3. | 1. | 40. | 45. | 1. | | *9. | | *40% $\Delta^9$ and 60% $\Delta^{11}$ | | c |
| | | 0.6 | 3.3 | 0.9 | 39.9 | 45.1 | 1.1 | tr. | 9.1 | | | 11 | d |
| 28 | | 2.97 | 4.05 | | 30.74 | 33.35 | | 0.36 | 3.56 | | ** | 12.2 | e |

**iso-13:0 = 1.58; iso-14:0 = 1.71; iso-16:0 = 2.16; 17:0 = 1.17; branched 17:1 = 0.36; others = 17.99.

a. Holton et al. (1964)
b. Parker et al. (1967)
c. Allen et al. (1966)
d. Hirayama (1967)
e. Oró et al. (1967)

composition was given as "others" and not further described.  It
is clear that Anacystis forms a quite reproducible fatty acid
pattern though it should be noted that all the cultures examined
are descendents of the original algal strain isolated about 15
years ago in the Austin, Texas area.

Although there is consistency of fatty acid composition within
a single algal species in different cultural conditions, examination of more than 20 different blue-green algae shows considerable
variation between species in the fatty acid compositional patterns
(Tables II and III).  One fatty acid that has never been found in
blue-green algae, although it appears to be present in all other
algae (except Cyanidium caldarium, Allen and Holton, 1969) and in
all green plants is trans-3-hexadececenoic acid (Nichols and James,
1968).  Its possible direct involvement in green plant photosynthesis
must be ruled out because blue-green algal photosynthesis has the
same mechanisms as far as we know now.

From a taxonomic point of view, it appears that there is some
direct correlation between the morphological complexity of the alga
and its fatty acid pattern (Holton et al., 1968).  Thus the only
organisms with mono-unsaturated acid pattern (Table II) are very
simple unicellular forms (Chroococcaceae) with the exception of
the Hapalosiphon.  Another isolate of this organisms (Mastigocladus,
a taxonomic synonym) has a small amount of diunsaturated acid
present, it should be noted.  This alga is found only in hot springs
and it has been suggested that such springs are very primitive
habitats and may contain "relic floras" which have changed very
little during recent biological evolution (Brock, 1967).

Another interesting observation concerns Spirulina platensis
in which  γ-linolenic acid rather than  α-linolenic is found
(Nichols and Wood, 1968).  γ-Linolenic acid is more typical of
"animal" metabolism than of plant systems, although in Spirulina
it is concentrated in the galactosyl diglycerides, which in chloroplasts and other algae is where the  α-linolenic acid is typically
localized.  Nichols and Wood suggest that Spirulina resembles the
marine algae in its fatty acid and lipid metabolism, a fact that
may be of some phylogenetic significance.  Other biochemical evidence such as that concerning the cytochrome composition suggests
that those algae with the simple fatty acid pattern like Anacystis
may indeed be the most evolutionarily primitive members of our
present flora and more closely related to the first autotrophic
cells on earth than any others we now have available to study
(Holton, 1969).

The question should be raised as to where in the cells the
fatty acids are localized and what their function is.  The first
question is the easier to try to answer.  The lipids from several
blue-green algae, including Anacystis and Anabaena variabilis,

## TABLE II

### FATTY ACID COMPOSITIONS OF BLUE-GREEN ALGAE WITH MONO- AND DIUNSATURATED ACID PATTERNS

| Family / Species | 12:0 | 14:0 | 14:1 | 15:0 | 16:0 | 16:1 | 16:2 | 18:0 | 18:1 | 18:2 | 18:3 | Others | Ref. |
|---|---|---|---|---|---|---|---|---|---|---|---|---|---|
| | Percentages (by weight of total) of Various Fatty Acids | | | | | | | | | | | | |
| | | | | Carbon atoms: Double bonds | | | | | | | | | |
| **Monounsaturated Pattern** | | | | | | | | | | | | | |
| Chroococcaceae | | | | | | | | | | | | | |
| Anacystis nidulans | tr. | 0.5 | 1.0 | | 47.0 | 38.8 | | 1.4 | 10.0 | | | 20:0=0.5 | a |
| Anacystis marina | tr. | 21% | 1.4 | 1.4 | 32. | 36. | | 1.6 | 4.1 | | | | b |
| Synechococcus cedrorum | 0.2 | 0.5 | 1.0 | | 47.0 | 38.8 | tr. | 1.4 | 10.0 | | | 17:1=0.5 | c |
| Mastigocladaceae | | | | | | | | | | | | | |
| Hapalosiphon laminosus | 0.4 | 1.0 | | | 53.7 | 23.8 | | 2.9 | 18.2 | | | 10:0=0.2 | c |
| **Diunsaturated Pattern** | | | | | | | | | | | | | |
| Oscillatoriaceae | | | | | | | | | | | | | |
| Oscillatoria williamsii | 0.6 | 2.0 | 1.2 | 1.7 | 36. | 24. | 14. | 1.9 | 11. | 3.9 | | | b |
| Lyngbya lagerhaimii | tr. | 2.0 | 0.8 | 1.1 | 40. | 15. | tr. | 1.9 | 31. | 7.4 | | | b |
| Nostocaceae | | | | | | | | | | | | | |
| Chlorogloea fritschii (heterotrophic) | tr. | 0.7 | tr. | | 39.2 | 19.0 | | 1.5 | 25.8 | 13.0 | 0.4 | 17:1=0.5 | c |
| Mastigocladaceae | | | | | | | | | | | | | |
| Mastigocladus laminosus | | | | | 38.5 | 42.5 | | tr. | 16.8 | 2.1 | | | d |

Note: Components may not total 100% in some algae as acids not listed on the table when present as less than 1.0% of the total have been omitted. In all tables, tr. = trace or less than 0.5%.

a. Holton et al., (1964)
b. Parker et al. (1967)
c. Holton et al., (1968)
d. Nichols and Wood (1968)

## TABLE III

### FATTY ACID COMPOSITIONS OF BLUE-GREEN ALGAE WITH TRIUNSATURATED ACID PATTERNS

| Family / Species | Percentages (by weight) of total of Various Fatty Acids | | | | | | |
| --- | --- | --- | --- | --- | --- | --- | --- |
| | \ Carbon atoms:Double bonds | | | | | | |
| | 12:0 | 14:0 | 14:1 | 15:0 | 16:0 | 16:1 | 16:2 |
| **Chroococcaceae** | | | | | | | |
| Agmenellum quadruplicatum | tr. | 2.0 | 1.2 | 1.8 | 34. | 15. | 3.8 |
| Coccochloris elabens | tr. | 1.1 | 0.5 | 0.8 | 49. | 12. | |
| Myxosarcina chroococcoides | | | | | 38.2 | 8.6 | 1.2 |
| **Oscillatoriaceae** | | | | | | | |
| Spirulina platensis | 2.2 | 21. | tr. | 1.4 | 43.4 | 9.7 | tr. |
| Trichodesmium erythaeum | 1.5 | 5.0 | tr. | 2.5 | 17. | 3.7 | |
| Microcoleus chthonoplastes | 0.2 | 1.3 | | | 37. | 13. | |
| Oscillatoria sp. | | | | | 29.0 | 24.0 | |
| O. chalybea (cells) | | tr. | | | 31. | 17.3 | |
| O. chalybea (thylakoids) | | tr. | | | 46.1 | 12.1 | |
| **Nostocaceae** | | | | | | | |
| Nostoc muscorum G | tr. | 2.7 | 2.1 | 2.2 | 27. | 20. | |
| Nostoc muscorum A | 1.0 | 1.6 | 2.0 | | 31.5 | 15.0 | 2.0 |
| Chlorogloea fritschii | 0.2 | 0.6 | 0.3 | | 41.2 | 16.5 | |
| Chlorogloea fritschii | tr. | 1.8 | 1.1 | 0.9 | 42.3 | 4.9 | tr. |
| Anabaena variabilis | | | | | 32. | 15. | |
| Anabaena flos-aquae | | | | | 39.5 | 5.5 | 4.3 |
| Anabaena cylindrica | | | | | 46.0 | 6.4 | 5.6 |
| **Scytonemataceae** | | | | | | | |
| Plectonema terebrans | tr. | 1.4 | 0.8 | 1.1 | 35. | 13. | 5.1 |

Table III.  Cont.

| Family / Species | Percentages (by weight) of total of Various Fatty Acids | | | | | Reference |
|---|---|---|---|---|---|---|
| | 18:0 | 18:1 | 18:2 | 18:3 | Others |  |
| | | | Carbon atoms:Double bonds | | |  |
| **Chroococcaceae** | | | | | | |
| Agmenellum quadruplicatum | 2.6 | 16. | 14. | 5.2 | | a |
| Coccochloris elabens | 1.2 | 13. | 17. | 2.6 | | a |
| Myxosarcina chroococcoides | 4.0 | 6.8 | 9.2 | 33.3 | | b |
| **Oscillatoriaceae** | | | | | | |
| Spirulina platensis | 2.9 | 5.0 | 12.4 | 21.4($\gamma$) | | b |
| Trichodesmium erythraeum | 2.6 | 2.8 | 4.2 | 19. | 10.0=27. | a |
| Microcoleus chthonoplastes | 3.7 | 14. | 5.0 | 18. | | a |
| Oscillatoria sp. | 1.5 | 25.6 | 10.2 | 6.8 | 17.0=1.5 | c |
| O. chalybea (cells) | 1.4 | 4.5 | 7.3 | 38.1 | | d |
| O. chalybea (thylakoids) | 2.4 | 4.7 | 5.9 | 27.9 | 21:0=0.9 | d |
| **Nostocaceae** | | | | | | |
| Nostoc muscorum G | 3.1 | 16. | 14. | 11. | | a |
| Nostoc muscorum A | 2.0 | 7.4 | 10.1 | 21.1 | 17:1=6.0 | c |
| Chlorogloea fritschii | 1.8 | 13.9 | 13.01 | 12.0 | 17:1=0.6 | c |
| Chlorogloea fritschii | 5.4 | 14.3 | 17.2 | 15.8 | | b |
| Anabaena variabilis | 4.4 | 14. | 14. | 17. | | a |
| Anabaena flos-aquae | 1.0 | 5.2 | 36.5 | 10.7 | | b |
| Anabaena cylindrica | 3.6 | 6.0 | 24.0 | 11.2 | | b |
| **Scytonemataceae** | | | | | | |
| Plectonema terebrans | 2.5 | 20. | 11. | 6.0 | | a |

a. Parker et al. (1967)
b. Nichols and Wood (1968)
c. Holton et al. (1968)
d. Schmitz (1967)

have been isolated and the fatty acids in each lipid determined.
The four predominant lipids are the same as those found in all
higher plant chloroplasts and in higher algae. They are the four
acyl lipids: the mono- and digalactosyl diglycerides, phosphatidyl
glycerol, and sulfolipid (Nichols et al., 1965; Allen et al., 1966;
Hirayama, 1967). Other neutral glycerides, phospholipids, and
glycolipids present in the photosynthetic tissues of higher plants
are not found in the blue-greens. Presumably these four lipids are
associated with proteins in the photosynthetic lamellae of the
blue-green algae and the chloroplasts. A morphological uniqueness
in the blue-green algae is that they lack the limiting membrane
bounding the chloroplast. Whether the lipids found in chloroplasts
but not in blue-greens are localized in the chloroplast membrane
is unknown. In those blue-green algae with polyunsaturated acids,
these acids are concentrated in the two galactosyl diglycerides
but in Anacystis, which lacks such acids, there is no particular
concentration of the mono-unsaturated acids in those lipids.
Because various changes in the environmental conditions such as the
growth temperature alter the fatty acid composition, one is led
to the conclusion that the exact nature of the fatty acid mixture
present is not of extreme importance to the structure of function
of alga. Nichols and James (1968) have summarized what little is
known about the possible structural and metabolic functions of
chloroplast lipids.

Recently studies with the blue-green alga, Chlorogloea fritschii
have been completed. It has several morphological states, unusual
for the blue-green algae. The stages labelled 1 through 4 are shown
in Figure 1. The small short filaments are described as Stage 1
and the large round cells as Stage 4. An examination of light and
temperature parameters has made it possible to discover how to
cause this alga to be predominately in one stage or another. During
growth, Stages II and III are intermediate in size and make up only
a small fraction of the total cell population (Findley, 1969).

Determination of the fatty acid content of Chlorogloea cells
grown under different conditions was carried out as follows: One
gram of dry, lyophilized cells was extracted by stirring with 20
mls. of 2:1 chloroform: methanol (v/v) for three hours (Nichols
et al., 1965). After filtration, the solvent was removed and the
extracted lipids were saponified by refluxing them in 5% KOH in
methanol for two hours (Holton et al., 1964). Non-saponifiable
material was extracted with ether and the basic solution was acidified and the free fatty acids were extracted with benezene. They
were esterified with 14% solution of $BF_3$ in methanol (Metcalfe and
Schmitz, 1961). The methyl esters were extracted with hexane and
separated on a 15% diethyleneglycol succinate polyester on Anachrom
ABS column at $176^{\circ}C$. in a gas chromatograph. A thermal conductivity
detector was used and identification of the acids was by comparison

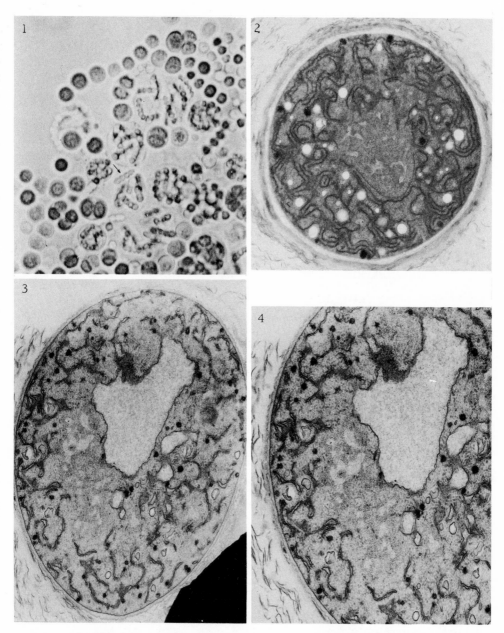

Figs. 1-4.  Fig. 1:  Mixed living culture of <u>Chlorogloea fritschii</u> showing
Stage I filaments (arrow) and large Stage IV cells.  Note Stage IV cells
forming in several Stage I cells.  X 1620.  Fig. 2:  Electron micrograph
of cells of <u>C. fritschii</u> grown at 45° C. and 25 ft. c.  Fixed in KMnO₄.
X 16,800.  Fig. 3:  Electron micrograph of cells grown at 45° C. and
700 ft. c.  Fixed in KMnO₄.  X 16,800.  Fig. 4:  Same as Fig. 3 but at
higher magnification.  X 23,600.

with carbon number vs. log retention time plots using known esters.
Separate portions of the methyl esters were reduced with hydrogen
gas using a PtO$_2$ catalyst (Farquher et al., 1959) to aid in identification of the unsaturated acids.

    The results of the fatty acid analyses are presented in Table
IV.  Although variations in light and temperature are large, the
differences in fatty acid composition are not great.  The big
differences seem to be due to the temperature effect in which the
unsaturation goes down and the chain length increases as the temperature is increased.  Most interesting is the big differences
in the amount of fatty acids present in different conditions.  It
would appear that Stage I cells, the short filaments, have much
higher acid content than do the Stage IV cells although the possible
significance of this observation must await further studies.

    In a related study of Chlorogloea (Findley, Walne, and Holton,
1969), electron microscopy has been used to study the ultrastruc-
tural of Chlorogloea as it is affected by light and temperature.
Figs. 2-4 show clearly that there are significant differneces at
the subcellular level of Stage IV cells grown at different light
intensities at the high temperature of 45°C.  In low light, the
lamellae are compact (Fig. 2), and the individual thylakoids are
long and continuous.  But in high light at the same temperature,
we find fewer lamellae and those present are fragmented with the
segments frequently swollen and vesiculate (Fig. 3-4).  And yet
the growth rates under both conditions are similar and the cells
appear healthy on superficial examination.  It would be valuable
to be able to relate the fatty acid analyses with the differences
in ultrastructure because a significant fraction of the lamellae
is lipid in nature.  However, it is not possible from our data to
make direct comparisons as the fatty acid analyses on the algae
when grown under the high light conditions contained mostly Stage
I cells while the electron microscopy was done on Stage IV cells
selected from this culture mixture.  But it does not appear that
as striking a change in acid composition has occurred as that
observable in the ultrastructure, although such changes might be
due to quantitative differences in lipid content rather than to
qualitative ones.

    Walsby and Nichols (1969) have separated the heterocysts, the
likely site of nitrogen fixation in blue-green algae, from the
vegetative cells of filaments of Anabaena cyclindrica.  They
report finding of a new lipophilic glycoside present but not found
in the vegetative cells.  While not presenting complete acid
analyses, they report that linolenic acid is not present in the
heterocysts although it is found in the vegatative cells.  The
four common acyl lipids are absent from the heterocysts which is
of interest as photosynthetic lamellae are not found in these cells.

## TABLE IV

FATTY ACID COMPOSITION OF CHLOROGLOEA FRITSCHII GROWN UNDER VARIOUS ENVIRONMENTAL CONDITIONS

| Light Intensity (ft. c.) | Temp. (°C.) | Percentages (by wt. of total) of fatty acids Carbon atoms:double bonds | | | | | | | Algal Growth Stage | % Fatty Acid Esters per Gram Dry Weight of Algae[a] |
|---|---|---|---|---|---|---|---|---|---|---|
| | | 14:0 | 16:0 | 16:1 | 18:0 | 18:1 | 18:2 | 18:3 | | |
| 25 | 25 | | 31 | 24 | | 17. | 18. | 10. | 90% Stage I | 2.8 |
| 500 | 25 | | 30. | 42. | | 21. | 7. | | 100% Stage IV | 0.02 |
| 700 | 45 | 1.4 | 31. | 19. | 6.4 | 37. | 5. | tr. | 85% Stage I | 0.2 |
| 25 | 45 | | 33. | 8. | 2. | 47. | 10. | | 100% Stage IV | 0.03 |
| dark | 35 | 0.7 | 39. | 19. | | 26. | 13. | 0.4 | 100% Stage IV | -- |

a  Percent of fatty acid esters is approximate and was determined from the composition of the sample from the gas chromatogram and relating this content back to the weight of the algal sample.

From fatty acid analyses done on blue-green algae, several
conclusions can be drawn.  Certain coccoid blue-green algae have
very simple fatty acid patterns; indeed the most simple of any
green plants, and there is not yet any obvious correlation of
fatty acid composition with function.  Some relationships can be
drawn between fatty acid patterns and morphological complexities
and these suggest that the development of biosynthetic abilities
is directly correlated with the evolution of morphological complexities.  Environmental conditions do have an effect on the
fatty acid compositions but there is not yet enough data available
to correlate these with structure.  Current fatty acid data suggest
that the blue-green algae on earth today, frequently considered
to be a small rather homogeneous group, may represent considerable
evolutionary diversity and its members may even include relic
organisms from very early periods of biological evolution.

## ACKNOWLEDGEMENTS

Previously unreported work at the University of Tennessee was
partially supported by National Science Foundation Grant GB-4203
to R.W.H.  The authors are indebted to Drs. Davis Findley and
Patricia Walne for the electron microscopy of Chlorogloea.

## REFERENCES

Allen, C.F. and R.W. Holton. 1969.  Lipids of Cyanidium caldarium.
    Unpublished manuscript.
Allen, C.F., O. Hirayama, and P. Good. 1966.  Lipid composition of
    photosynthetic systems.  In Biochemistry of Chloroplasts.
    (T.W. Goodwin, ed.), Academic Press, London and New York, 1,
    195-199.
Brock, T.D. 1967.  Life at high temperatures.  Science 158, 1012-
    1019.
Farquhar, J.W., W. Insull, Jr., P. Rosen, W. Stoffel, and E.H. Ahrens,
    Jr. 1959.  The analysis of fatty acid mixtures by gas-liquid
    chromatography:  Construction and operation of an ionization
    chamber instrument.  Nurt. Rev. 17, Suppl., 1-30.
Findley, D.L. 1969.  A study of the effect of environment variations
    on the growth development cycle, and ultrastructure of the
    blue-green alga Chlorogloea fritschii.  Ph.D. Dissertation,
    The University of Tennessee, Knoxville.
Findley, D.L., P.L. Walne, and R.W. Holton. 1969.  The effects of
    temperature and light intensity on the ultrastructure of
    Chlorogoea fritschii Mitra.  Unpublished manuscript.
Hirayama, O. 1967.  Lipids and lipoprotein complex in photosynthetic
    tissues. II. Pigments and Lipids in blue-green alga, Anacystis
    nidulans.  J. Biochem. (Tokoyo) 61, 179-185.

Holton, R.W. 1969. Blue-green algae, primitive cells, and compara-
    tive biochemistry. XI International Botanical Congress,
    Seattle, Washington, Abstracts, p. 93.
Holton, R.W., H.H. Blecker, and M. Onore. 1964. Effect of growth
    temperature on the fatty acid composition of a blue-green alga.
    Phytochemistry 3, 595-602.
Holton, R.W., H.H. Blecker, and T.S. Stevens. 1968. Fatty acids in
    blue-green algae: possible relation to phylogenetic position.
    Science 160, 545-547.
Levin, E., W.J. Lennarz, and K. Bloch. 1964. Occurrence and local-
    ization of α-linolenic acid containing galactolipids in the
    photosynthetic apparatus of Anabaena variabilis, Biochem.
    Biophys. Acta 84, 471-474.
Mazur, A. and H.T. Clarke. 1942. Chemical composition of some
    autotrophic organisms. J. Biol. Chem. 143, 39-42.
Metcalfe, L.D. and A.A. Schmitz. 1961. The rapid preparation of
    fatty acid esters for gas chromatographic analysis. Anal.
    Chem. 33, 363-364.
Nichols, B.W. and B.J.B. Wood. 1968. The occurrence and biosynthe-
    sis of gamma-linolenic acid in a blue-green alga, Spirulina
    platensis. Lipids 3, 46-50.
Nichols, B.W., and A.T. James. 1968. The function and metabolism
    of fatty acids and acyl lipids in chloroplasts. In Plant Cell
    Organelles (J.B. Pridham, ed.), Academic Press, London and
    New York, p. 163-178.
Nichols, B.W., R.V. Harris and A.T. James. 1965. The lipid meta-
    bolism of blue-green algae. Biochem Biophys. Res. Commun.
    20, 256-262.
Oro, J., T.G. Tornabene, D.W. Nooner, and E. Gelpi. 1967. Aliphatic
    hydrocarbons and fatty acids of some marine and freshwater
    microorganisms. J. Bacteriol. 93, 1811-1818.
Parker, P.L., C. Van Baalen, and L. Maurer. 1967. Fatty acids in
    eleven species of blue-green algae: geochemical significance.
    Science 155, 707-708.
Schmitz, R. 1967. Über die Zusammensetzung der pigmenthaltigen
    Strukturen aus Prokaryonten. I. Untersuchungen an Thylakoiden
    von Oscillatoria chalybea Kutz. Arch. Mikrobiol. 56, 225-237.
Walsby, A.E. and B.W. Nichols. 1969. Lipid composition of hetero-
    cysts. Nature 221, 673-674.

# ALGAL SULFOLIPIDS AND CHLOROSULFOLIPIDS

Thomas H. Haines

The City College of The City University of New York

New York, New York 10031

The occurrence in algae of large quantities of sulfolipids and chlorosulfolipids often dominates the sulfur and chloride requirements in these microbes. Until this last decade, however, such substances were unknown and naturally it was not expected that they could possibly occur in sufficient quantity to have any bearing on the nutritional requirements of algae.

In 1960 a study was initiated on the sulfur metabolism of algae (Haines and Block, 1962). As a first step in this study it was deemed necessary to incubate the algae with $S^{35}$-sulfate and conduct a paper chromatographic survey of the sulfur-containing compounds. Three algae were selected for this study, Chlorella pyrenoidosa, Ochromonas malhamensis, and O. danica. Cells were extracted with cold trichloracetic acid (TCA), ethanol:ether, hot TCA and the residue hydrolysed with acid to examine the protein bound amino acids. An autoradiogram of a paper chromatogram of the ethanol:ether extract of O. danica is shown in Figure 1. The auto-radiogram shows that large amounts of sulfolipid (unknown No. 1) are synthesized by the organism. The cold TCA extract also showed a large amount of this material, whereas the protein hydrolysate showed a relatively small amount of cystine and methionine. It was subsequently established that the sulfolipids constituted approximately 60 to 80% of the sulfur in the cell and 3% of the dry weight (Haines and Block, 1962; Elovson and Vagelos, 1969).

A report had just been published by Benson et al. (1959) describing a very novel sulfolipid in C. pyrenoidosa and higher plants. The substance is a sulfonolipid (sulfonic acid containing lipid) (Haines, 1970). The structure was subsequently shown by Daniel et al. (1961) to be diacyl-6-sulfo-O-α-D-quinovosly-(1 → 1)

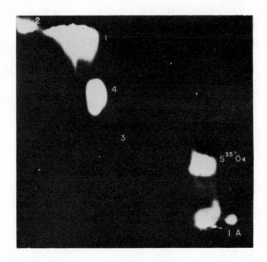

Fig. 1.    Autoradiogram of a paper chromatogram of a lipid
extract of O. danica grown in the presence of S$^{35}$-sulfate.    Solvents and conditions of chromatography have been described
(Haines, 1962).

glycerol:

$$SO_3^-$$

$$CH_2$$

HO — C — O — C — O — CH$_2$

H$_2$C — O — C — R

HC — O — C — R

H — C — C — $\underline{\alpha}$ H

OH    OH    H

OH    H

    This sulfolipid has been shown to occur in all chloroplasts
and there is some evidence for its participation in photosynthesis
(Haines, 1970).    On examination of C. pyrenoidosa a large quantity
of sulfolipid was found on the autoradiogram of a paper chromatogram of the lipid extract of this alga.    Co-chromatography of the
two extracts, however, established that the sulfolipids in O. danica were different from the sulfonolipid in Chlorella (Figure 2).

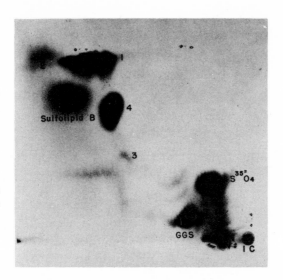

Fig. 2.    Autoradiogram of a paper chromatogram of lipid
extracts of O. danica and C. pyrenoidosa co-chromatographed.
Sulfolipid B is the plant sulfolipid reported by Benson et al.
(1959) and GGS is the glycerol glycoside sulfonate.

Chemical studies on the labelled sulfolipids soon revealed
that the substances were sulfate esters and not sulfonolipids.
It was also noted that the material is excreted into the medium
(Benson et al. 1959).  An autoradiogram of a paper chromatogram
of $S^{35}$-labelled medium of O. danica is shown in Figure 3.  It had
already been reported that sulfate esters were not a nutritional
form of sulfur in higher plants (Lowe and Delony, 1961).  To re-
examine this phenomena a batch of radioactively pure sulfatide was
purified and fed it back to the organism.  The microbe was unable
to cleave the sulfate ester since no $S^{35}$-cystine of $S^{35}$-methionine
was found in the cells.  It is quite possible that algae are
generally incapable of utilizing the sulfur of ester sulfates.

Chemical studies on the isolated sulfolipid showed it was a
mixture of substances.  Analysis of the mixture (Mayers and Haines,
1967) indicated a sulfur:carbon ratio of 1:11.  An infrared spectrum of the preparation demonstrated the presence of both primary
and secondary sulfate esters.  These data indicated a mixture of
$C_{22}$ disulfates.  In order to remove the sulfates gently and without disturbing the C-O bond a simple refluxing was established in
moist dioxane for 10-15 min.  This procedure quantitatively removes
the sulfate to produce the corresponding alcohol with complete

retention of the configuration (Mayers et al. 1969).

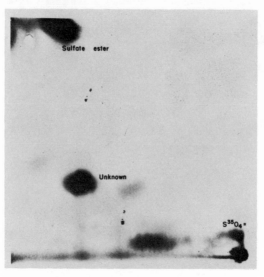

Fig. 3.    Autoradiogram of a paper chromatogram of the culture medium of C. pyrenoidosa after growth in the presence of $S^{35}$-sulfate. Autoradiograms of the media of O. danica and O. malhamensis were virtually identical.

This procedure produced a mixture of diols which were separated on thin layer chromatography as shown in Figure 4.   This plate is a recombined group not the original solvolysate.   The mixture was placed on a silicic acid column and fractions collected as seen on the thin layer plate.   Individual fractions were examined by elemental analysis, mass and nuclear magnet resonance spectrometry and infrared spectrophotometry.   The two lower spots at position 8 on the chromatogram were the first analysed (Mayer and Haines, 1967; Mayer et al, 1969).   The mass spectrum of the principal component (the lower spot) clearly demonstrated that it is 1,14-docosanediol.   This was in confirmation of the other spectral and analytical evidence.   Synthesis of racemic 1,14-docosanediol and examination of the synthetic material confirmed the assignment. Figure 5 shows a gas chromatograph (GC) of the fraction; less than 10% of the mixture is the tetracosanediol.   Using a GC-mass spectrometer hook-up Elovson and Vagelos (1969) showed that the secondary hydroxyl on this compound was in the 15-position.

Also noted were the well known and very characteristic isotopic chloride patterns obtained from the mass spectra of polychloro compounds.   Thus it was shown that the diols contain chlorides replacing hydrogen atoms on the diol chains.   This suggested that the next pair of compounds at position 7 on the TLC plate

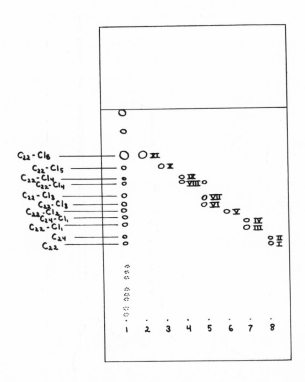

Fig. 4.   Thin layer chromatogram of the diols obtained from the sulfolipids of <u>O. danica</u> by solvolysis in dioxane (1) (Mayer et al, 1969), and of the fractions obtained from a column of silicic acid (2-8).  Roman numerals refer to the structures in Schemes II and III.  Procedures for thin layer and column chromatography have been described.

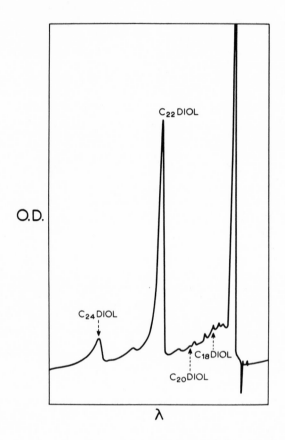

Fig. 5.    Gas chromatogram of natural 1,14-docosanediol and 1,15 tetracosanediol obtained by solvolysis of the sulfatide in dioxane. The positions of the hydroxyls in the $C_{18}$ and $C_{20}$ diols are not known. The column, consisting of 3% JXR on Chromosorb Q, was maintained at 230°.

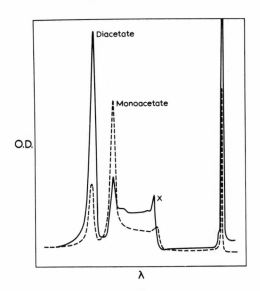

Fig. 6.    G.l.c. of chlorodocosanediol after acetylation for
30 min. (......) and 1 hr. (———) with acetic anhydride in
pyridine.  Samples were injected directly into the Perkin-Elmer
model 881 chromatograph containing a 6 ft. x 1/8 in. column (3%
JXR on Chromosorb Q) at 230°.

(Figure 4) should be analysed (Haines et al, 1969) for its chloride content.  An analysis showed it to contain 1 chlorine atom
for 22 carbons and 2 oxygens.  Since difficulty was encountered
in the gas chromatography of this material, it was necessary to
follow by acetylation and determination by GC (Figure 6).  Without acetylation it decomposed to a compound (X) which turned out
to be 14-ketodocosanol judging from its mass spectrum (Haines
et al, 1969).  The acetylation study showed, however, that decom-
position was prevented by acetylation and that indeed it was a
diol.  The mass spectrum of this substance (Figure 7) showed it
to be 13-chloro-1,14-docosanediol.  Thus the compound was a
chlorohydrin.

As a chlorohydrin it would be expected to form an epoxide in
base.  Furthermore the formation of an epoxide would be stereo-
specific and allow us to identify the configuration as three or
erythro.  Thus a three chlorohydrin forms a cis epoxide and an
erythro chlorohydrin forms a trans epoxide (scheme I).  Fortunately

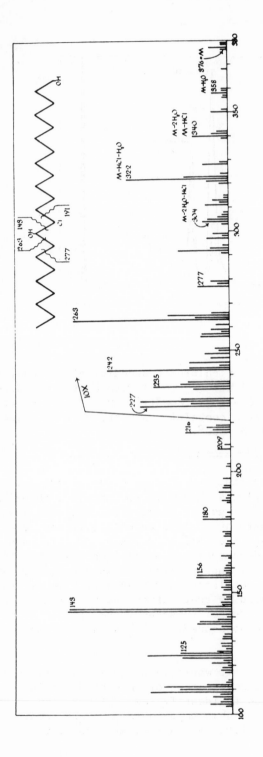

Fig. 7.   Mass spectrum of 13-R-chloro-1-(R)-14-docosanediol isolated from O. danica.

Roomi, et al. (1966) had shown that these isomeric epoxides are easily separated and thereby identified by TLC. Therefore, the cis and trans 13-epoxy-docosanols were synthesized from erucic and brassidic acids respectively. The epoxide derived from the natural chlorodiol co-chromatographed with the cis-13-epoxy-docosanol which demonstrated that the compound was threo.

### Schemes

The rotations of long chain chlorohydrins have been studied by Morris and Wharry (1966). The rotation of the chlorodiol was $|\alpha|_D^{27°} = +14.7$ (C, 0.0336 in chloroform) and comparison of this rotation with the data of Morris and Wharry (1966) confirmed the threo assignment and established that the substance is threo-(R)-13-chloro-1-(R)-14-docosane diol. The configuration of the chloro is L and that of the hydroxyl is D. The configuration of the 1,14-docosanediol (without the chloro group) had been found to be L (Mayers et al, 1969). Based on the mass spectrum of the trimethyl silyl ether of this substance Elovson and Vagelos (1969) have also found it to be 13-chloro-1,14-docosanediol.

Additionally they have confirmed our assignment of the 1,14-docosanediol and found the front running spot of Fraction 7 to be 14-chloro-1-15-tetracosanediol. This is in agreement with our data which suggest that the substance is threo-(R)-14-Chloro-1-(R)-15-tetracosanediol. Likewise the structure of the compound at position 6 on the TLC plate (Figure 4) appears to be erythro-11-(S)-15-dichloro-1-(R)-14-docosanediol. These data are summarized in Scheme II.

The remaining substances are isomers of 1,14-docosanediol with increasing numbers of chlorines replacing the hydrogens up to a maximum of six. The precise location of the chloro groups on the chain has not been established although the mass spectra clearly indicate how many chloro groups are on each side of the 14-hydroxyl as indicated. Apparently six is the maximum number of chlorogroups that can be placed on the chain. The hexachloro compound dominates the mixture along with the "a-chloro" compound. The other substances remain relatively low in quantity. It should be recalled that these substances are disulfates and not diols in the algae.

An organism has been cultured - which is normally a fresh water phytoflagellate - in 3/8 saline concentration (0.72% chloride) which is just about the maximum salt it can tolerate. Under these conditions the organisms grows much slower than in its normal medium. The proportion of hexachlorosulfolipid increases with increasing salt concentration in the growth medium. At maximum salt concentration the "a-chloro" docosane disulfate is still present and at

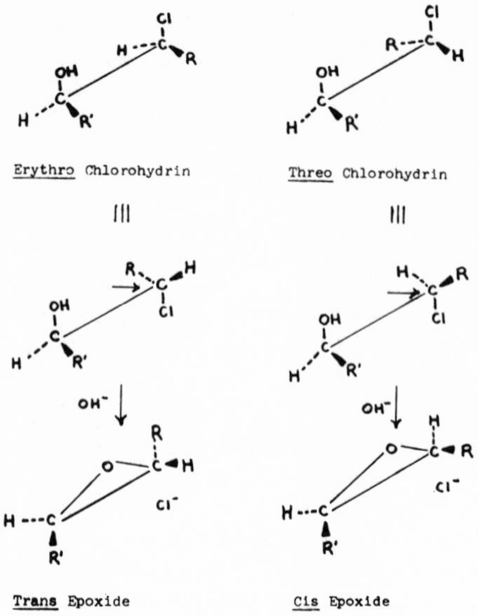

Scheme I

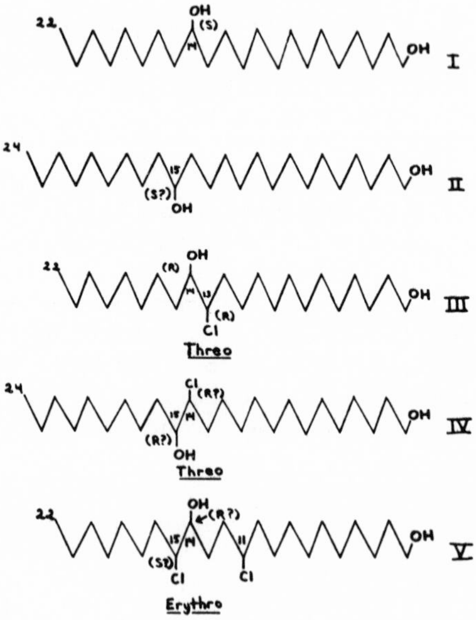

Scheme II

a minimal chloride concentration the hexachlorodocosane disulfate
is present.  The minimum tolerable chloride concentration has not
het been tested - the low chloride concentration mentioned here
is that in which chloride is only added as the hydrochloride of
the nutritive amino acids that were included in the defined medium.
It is interesting that the chlorination of the docosyl disulfate
is proportional to chloride concentration in the medium but the
available information does not suggest how the organism controls
the chlorination nor why it fixes chloride at all.  Until the
discovery of these compounds it had always been considered that
the amount of chloride in the medium was only required for its
osmotic effects.  In the future its biochemical activity must be
accounted for in algal culture.

All sulfolipids have been demonstrated to be constituents of
membranes where their location in the cell has been explored.  The
only other sulfolipid in algae, the chloroplast sulfonolipid, has
been shown in several laboratories to be a constituent of the
chloroplast membrane (Haines, 1970).  The sulfatide described herein is probably in a particulate fraction of the cell and a part of
the plasma membrane.

Attempts to explore the biosynthesis of this substance have
also been made.  $C^{14}$-acetate and $C^{14}$-octanoate are effective
precursors for the sulfolipid complex, whereas $C^{14}$-laurate and
$C^{14}$-oleate are not incorporated into these substances.  Clearly
the 8-carbon chain of octanoate can be incorporated as is into the
docosyl disulfate but a 12-carbon chain is lacking an appropriate
functional group.  These data suggest that the synthesis of the
sulfolipids and chloro sulfolipids require special enzymes which
conduct reactions on substrates of the $C_{10}$ to $C_{12}$ range.  A care-
ful examination of the structures of the various known sulfatides
and chlorosulfatides (Scheme II) suggests the biosynthetic route
of Scheme IV.  The route suggests that 2-<u>trans</u>-dodecenoyl ACP (acyl
carrier protein) is converted by a specific isomerase to either the
3-hydroxy derivative or the 3-<u>cis</u> compound.  This suggestion is
based on the occurrence of such an enzyme in <u>Escherichia coli</u>
(Helmkamp et al, 1968).  The 3-hydroxy-dodecanoyl-ACP would be a
direct precursor for 1,15-tetracosyl disulfate.  A second isomerase
of the same type is postulated which allows for the formation of
4-hydroxy-dodecanoyl-ACP (1,14-docosanyl disulfate) and the 4-<u>trans</u>-
dodecenoyl thio ester.  The biosynthetic route assumes that the
chlorohydrins that are formed all occur with the same type of
enzyme synthesis - namely a <u>trans</u> addition of oxygen and chlorine
to the double bond.  Thus the epoxide is first formed and chloride
opens the epoxide ring.  An important alternative is the formation
of Cl+ which is brought into contact with the double bond and
neutralized with hydroxide.

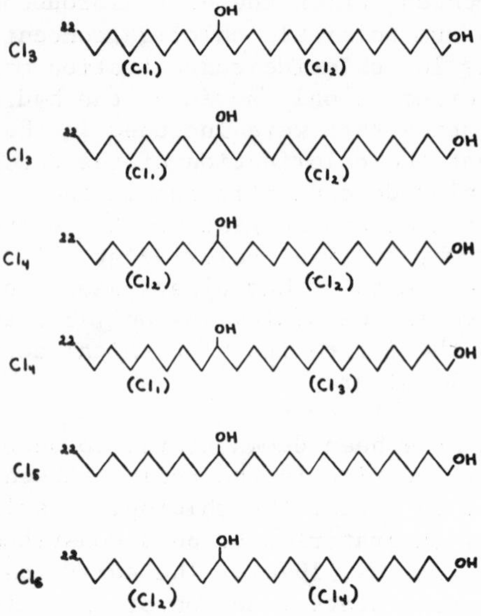

Cl$_3$ 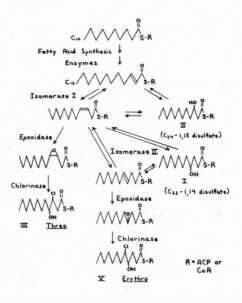 OH ...OH  **VI**
(Cl$_1$)   (Cl$_2$)

Cl$_3$  OH  ...OH  **VII**
(Cl$_1$)   (Cl$_2$)

Cl$_4$  OH  ...OH  **VIII**
(Cl$_2$)   (Cl$_2$)

Cl$_4$  OH  ...OH  **IX**
(Cl$_1$)   (Cl$_3$)

Cl$_5$  OH  ...OH  **X**

Cl$_6$  OH  ...OH  **XI**
(Cl$_2$)   (Cl$_4$)

Scheme III

Scheme IV

This route is attractive because it explains the large
variety of compounds with a minimal number of enzymes.  At present
however, the scheme remains speculation.

In summary, a series of 1,14-docosane disulfates and 1,15-
tetracosanedisulfates have been found in Ochromonas danica and
related algae.  The substances contain up to six chloro groups
substituting for hydrogens on the chain.  They constitute a major
portion of the substance of O. danica cells (60 to 80% of that is
fixed by the cells varies with its concentration in the growth
medium.  A large quantity of these substances are excreted by the
cells.

Acknowledgements:  The author wishes to express his gratitude
to Dr. G.L. Mayer, M. Pousada, B. Stern, R. Orner, C. Rapoport,
N. Rosenbaum and C. Mooney for their contributions toward the
development of these data.  This work was supported by The Department of Interior, Water Pollution Control Administration (WP-00675)
and by the National Science Foundation Undergraduate Research
Participation Program.  Particular appreciation is expressed to
Mrs. Rose Gruber for assistance in the preparation of the manu-
script.

## References

Benson, A.A., H. Daniel, and R. Wiser. 1959.  A sulfolipid in
      plants.  Proc. Natl. Acad. Sci. 45, 1582-1587.
Daniel, H., M. Miyano, R.O. Mumma, T. Yagi, M. Lepage, I. Shibuya
      and A.A. Benson. 1961.  The plant sulfolipid identification
      of 6-sulfo-quinovose.  J. Amer. Chem. Soc. 83, 1765.
Elovson, J. and P.R. Vagelos. 1969.  A new class of lipids:
      chlorosulfolipids.  Proc. Natl. Acad. Sci. 62, 957-963.
Haines, T.H. and R.J. Block. 1962.  The sulfur metabolism of algae.
      I. Synthesis of metabolically inert chloroform-soluble sulfate
      esters by two chrysomonads and Chlorella pyrenoidosa.  J.
      Protozool. 9, 33-38.
Haines, T.H. 1970. in Progress in the Chemistry of Fats and Other
      Lipids.  R.T. Holman, Ed., Pergamon Press, N.Y.
Haines, T.H., M. Pousada, B. Stern and G.L. Mayers, 1969.  Micro-
      bial sulpholipids: (R)-13-chloro-1-(R)-14-docosanediol
      disulphate and polychlorosulpholipids in Ochromonas danica.
      Biochem. J. 113, 565-566.
Haines, T.H. 1965.  The biochemistry of the sulfolipids.  J.
      Protozool. 12, 655-659.
Helmkamp. G.M. Jr., R.R. Rando, D.J.H. Brock and K. Block. 1968.
      β-hydroxydecanoyl thioester dehydrase.  Specificity of sub-
      strates and acetylenic inhibitors.  J. Biol. Chem. 243,
      3229-3235.

Lowe, L.E. and W.A. Delong. 1961. Aspects of the sulfur status
    of three Quebec soils. Can. J. Soil. Sci. 41, 141-146.
Mayers, G.L., M. Pousada and T.H. Haines. 1969. Microbial sulfo-
    lipids. III. the disulfate of (+)-1,14-Docosanediol in
    Ochromonas danica. Biochemistry 8, 2981-2986.
Mayers, G.L. and T.H. Haines. 1967. A microbial sulfolipid. II.
    Structural studies. Biochem. 6, 1665-1671.
Miyano, M. and A.A. Benson. 1962. The plant sulfolipid. VI and
    VII: Configuration of the glycerol moiety; Synthesis of
    6-sulfo-α-D-quinovopyranosyl glycerol and radiochemical
    synthesis of sulfolipids. J. Amer. Chem. Soc. 84, 57-59,
    59-62.
Morris, L.J. and D.M. Wharry. 1966. Naturally occurring epoxy
    acids. IV. The absolute optical configuration of vernolic
    acid. Lipids 1, 41-46.
Okaya, Y. 1964. The plant sulfolipid: a crystallographic study.
    Acta Crystallog. 17, 1276-1282.
Orner, R. and T.H. Haines. 1969. The occurrence of 1,14-doco-
    sanediol disulfate in Ochromonas membrane. Unpublished
    observations.
Roomi, M.W., M.R. Subbaram and K.T. Achaya. 1966. Separation of
    epoxy, hydroxy, halohydroxy, keto fatty acids and derivatives
    by thin layer chromatography. J. Chromatography 24, 93-98.